# 8th International Conference on Highly Frustrated Magnetism (HFM 2016)



Journal of Physics: Conference Series Volume 828

Taipei, Taiwan
7 – 11 September 2016

ISBN: 978-1-5108-3981-6
ISSN: 1742-6588

**Printed from e-media with permission by:**

Curran Associates, Inc.
57 Morehouse Lane
Red Hook, NY 12571

Some format issues inherent in the e-media version may also appear in this print version.



Printed by Curran Associates, Inc. (2017)

For permission requests, please contact the Institute of Physics
at the address below.

Institute of Physics
Dirac House, Temple Back
Bristol BS1 6BE UK

Phone: 44 1 17 929 7481
Fax: 44 1 17 920 0979

techtracking@iop.org

**Additional copies of this publication are available from:**

Curran Associates, Inc.
57 Morehouse Lane
Red Hook, NY 12571 USA
Phone: 845-758-0400
Fax: 845-758-2633
Email: curran@proceedings.com
Web: www.proceedings.com

# Table of contents

## Volume 828

**8th International Conference on Highly Frustrated Magnetism 2016**
**7–11 September 2016, Taipei, Taiwan**

**Accepted papers received: 21 March 2017**
**Published online: 20 April 2017**

IOP Conf. Series: Journal of Physics: Conf. Series **828** (2017) 011001 doi:10.1088/1742-6596/828/1/011001

# The 8th International Conference on Highly Frustrated Magnetism (HFM 2016)

**J. S. Gardner**

Neutron Scattering Group, NSRRC, Hsinchu, 30076 Taiwan and
National Taiwan University, Taipei, 10607 Taiwan

**Y. J. Kao**

Department of Physics, National Taiwan University, No. 1, Sec. 4, Roosevelt Rd. Taipei, 10607, Taiwan

E-mail: gardnerj@nsrrc.org.tw

**Abstract.**

The 8th International Conference on Highly Frustrated Magnetism 2016 (HFM 2016) took place between the 7th and 11th of September 2016 at the GIS Convention Center at National Taiwan University, Taipei, Taiwan. Over 260 participants from all over the world, attended the meeting making it the largest HFM to-date and revealing the impressive growth in the community since the original meeting in Waterloo, Canada where 80 participants attended. Preceding the meeting a school was held at the National Synchrotron Radiation Research Center to help those new to the field understand the material they were likely to see at HFM2016. Our thanks to the international speakers who attended this school John Chalker, Michel Kenzelmann, Philippe Mendels, Luigi Paolasini, Kirrily Rule, Yixi Su, Isao Watanabe and those from Taiwan W. T. Chen, Y-J, Kao, L. J. Chang and C. S. Ku, for their enlightening presentations.

The HFM 2016 conference consisted of five plenary talks by H Takagi, B D Gaulin, L Balents, Y Tokura and S T Bramwell, 20 invited and 40 contributed presentations, and about 160 poster presentations from all aspects of theoretical and experimental frustrated magnetism. During the conference period, many stimulating discussions were held both inside and outside the conference room. Excursions to Taipei 101 and the National Palace Museum, as well as several organized dinners and receptions allowed the participants to initiate collaborations and discuss the hottest issues. The subjects covered in the conference included:

· Quantum frustrated magnetism and spin liquids
· Novel ordering of geometrically frustrated magnets
· Frustration effect on the coupling to lattice, orbital and charge degrees of freedom
· Exotic phenomena induced by macroscopic degeneracy
· Field effect on frustrated magnetism etc.

These proceeding represent a very small, but valuable contribution to the community. I hope you enjoy reading them. In view of the rapid growth of the field, it has been decided at the conference that the next HFM conference will be held in two years, in 2018, on the campus of UC Davis, followed one in Shanghai around December 2019.

Finally, on behalf of the HFM 2016 Organizing Committee, I wish to deeply thank all the people who generously helped us in organizing and running the HFM 2016 conference. These include the numerous funding sources, Committee Members, Program and Session chairs, Chia-Chi Liu, Chao-Jung Kuo, Hanz Peng, Laura Bravo and Stella Su and of course all the participants, without you it would have simply been another wet week in Taipei.

One might remember HFM2016 for the power outage, but we hope you remember the taste, sounds and views of Taipei as well as stimulation conversations and hopefully the beginnings of productive partnerships.

# Peer review statement

All papers published in this volume of *Journal of Physics: Conference Series* have been peer reviewed through processes administered by the proceedings Editors. Reviews were conducted by expert referees to the professional and scientific standards expected of a proceedings journal published by IOP Publishing.

IOP Conf. Series: Journal of Physics: Conf. Series **828** (2017) 012001 doi:10.1088/1742-6596/828/1/012001

# Dielectric Property and Diffuse Scattering in Relaxor Magnet $LuFeCoO_4$

**Minoru Soda and Takatsugu Masuda**

Neutron Science Laboratory, Institute for Solid State Physics, University of Tokyo, Tokai, Ibaraki 319-1106, Japan

E-mail: soda@issp.u-tokyo.ac.jp

**Abstract.** Dielectric permittivity and nuclear diffuse scattering were measured on a single crystal of relaxor magnet $LuFeCoO_4$ having the triangular lattice. The dielectric constant with the electric field applied in the $c$-plane is much larger than that with the electric field applied along the $c$-axis. We found that the polar nanoregion (PNR) has the anisotropy, and that the strong coupling between PNR and magnetic ordering exists.

Novel relationships between magnetism and dielectricity are attracting great interest in the fields of materials science and condensed matter physics [1]. A number of studies had revealed its mechanism in multiferroics [2]. In order to find other system exhibiting the correlation between magnetism and dielectricity, we focus on a relaxor ferroelectrics having magnetic ions, relaxor magnet, such as a complex perovskite (1-$x$)$BiFeO_3$-$x$$BaTiO_3$ [3]. In relaxor ferroelectric systems [4], temperature dependence of the dielectric permittivity shows a broad maximum, and both value and the peak temperature have a frequency dependence. One of the most important concept of relaxor system is “Polar Nanoregions” (PNRs), where ordered polarizations in nanoscale domains are randomly oriented. Physical properties of many relaxor ferroelectrics have been understood based on the behavior of PNRs.

The relationships between relaxor property and magnetism have been rarely reported because the typical relaxor systems have no magnetic ions. Previously we studied one of the relaxor magnets $LuFeCoO_4$ [5]. $LuFeCoO_4$ has the triangular lattice and the $Fe^{3+}$ and $Co^{2+}$ ions are allocated randomly on the same site. The triangular lattice formed by the $Fe^{3+}$ and $Co^{2+}$ ions has the geometrical spin and valence frustrations. $LuFeCoO_4$ also has a relaxor behavior in the dielectric properties [6]. Thus, we can conclude that this system is the relaxor magnet. In our elastic neutron study of $LuFeCoO_4$, the strong coupling between the nuclear and magnetic correlations has been found. The simultaneous change in the nuclear and magnetic diffuse intensities was observed at $T \sim 190$ K. Furthermore, the magnetic correlation length along (1,1,0) direction at 190 K coincides the size of the PNRs estimated from nuclear diffuse scattering in the $c$-plane. From these neutron results, we propose “multiferroic nano region” (MNR) model [3]. At 190 K the correlation of both magnetic moments and electric polarizations are developed inside the MNRs. Outside the MNRs the nuclear correlation is absent and so is the magnetic correlation. With decreasing the temperature at $T \sim 190$ K, the magnetic domain size increases gradually. In the temperature region 70 K $< T <$ 190 K, spin moments having ferromagnetic component in nano-magnetic domain behave as superparamagnetic moments. In the present study, we studied the anisotropies of the dielectric property and nuclear diffuse scattering in

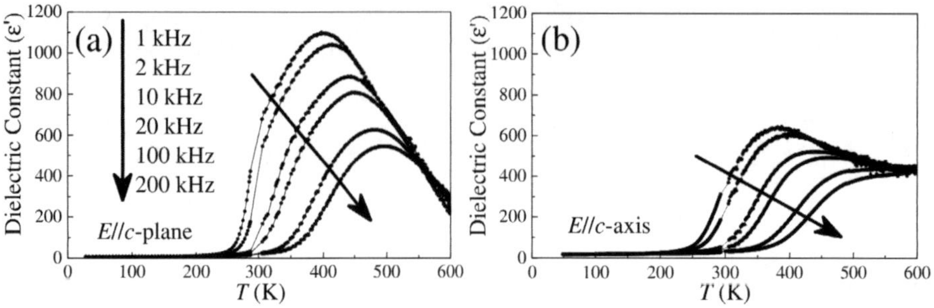


**Figure 1.** Temperature dependence of dielectric constant measured at various frequencies with the electric field applied (a) in $c$-plane and (b) along $c$-axis.

the relaxor magnet $LuFeCoO_4$. Our results suggests that the anisotropic PNR is realized in the triangular lattice $LuFeCoO_4$.

Single crystal of $LuFeCoO_4$ was grown by a floating zone method. The obtained crystal was annealed at 600 °C for 120 hours in oxygen gas. The dielectric permittivities at low and high temperatures were measured with an $LCR$ meter using a closed cycle refrigerator and an infrared heating furnace, respectively. A neutron diffraction experiment was carried out using the triple-axis spectrometer HQR installed at JRR3 in JAEA Japan. The crystal was oriented so that the $hhl$-plane was the scattering plane; throughout this paper, we use the hexagonal unit cell with $a = b = 3.42$ Å and $c = 25.28$ Å.

Figures 1(a) and 1(b) show the temperature dependences of the real part of the dielectric constant $\epsilon'$ measured at various frequencies with the electric field $E$ applied in the $c$-plane and along the $c$-axis, respectively. The large dispersions of the dielectric permittivities, which were induced by the valence frustration of $Fe^{3+}$ and $Co^{2+}$ ions [7], were observed. For both $E$-directions the peak of $\epsilon'$ shifts progressively toward higher temperatures, and the maximum value of the $\epsilon'$ decreases with increasing the frequency. However, the dielectric constant with $E$ applied in the $c$-plane is much larger than that with $E$ applied along the $c$-axis. This means that the polar direction fluctuates mainly in the $c$-plane.

Figures 2(a) shows the intensity distribution of the nuclear scattering around the (1,1,0) Bragg reflection at 300 K. Nuclear diffuse scattering was clearly observed and it extends approximately along the (0.05,0.05,±1) direction. Although there was an imbalance between the clusters flattened perpendicular to (0.05,0.05,1) and (0.05,0.05,−1) directions in the previous study [3], the clusters in the present crystal were almost equal in population. Then, so-called butterfly-shaped diffuse scattering is observed as reported in conventional relaxor [8].

Figure 2(b) shows the contour plot of the neutron intensity around the (1,1,0) Bragg reflection at 600 K. With increasing temperature the intensity of the nuclear diffuse scattering decreases and the shape approaches to be isotropic. The temperature region where the strong nuclear diffuse scattering is observed coincides with the temperature region where the dielectric permittivity is large. This means that the nuclear diffuse scattering is induced by PNR, as reported in the previous study [3]. The number of PNR which induces the nuclear diffuse scattering decreases with increasing temperature.

Here we discuss the anisotropy of PNR from the dielectric permittivity and the nuclear diffuse scattering. The temperature dependence of the dielectric permittivity means that the polar direction in PNR fluctuates mainly in the $c$-plane. Two-dimensionality of the triangular lattice, the $Lu^{3+}$ ion layer or the transition-metal layer formed by $Fe^{3+}$ and $Co^{2+}$ ions, induces the anisotropy of PNR. The in-plane flexibility of the polar direction in the triangular lattice is expected to allow the larger two-dimensionality of the crystal structure because the polar

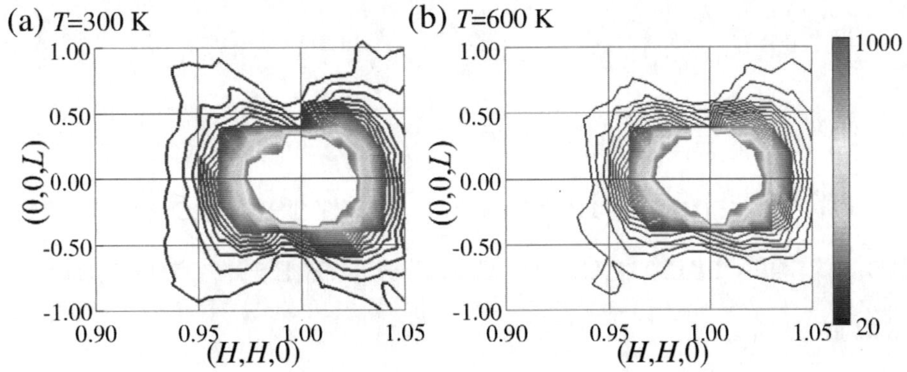


**Figure 2.** (Color online) Contour plots of the nuclear diffuse intensity at (a) 300 K and (b) 600 K around the 110 reflection.

directions of PNRs in the neighbor triangular lattices are independent. Then, the nuclear diffuse scattering extends even along $c$-axis. With decreasing temperature the nuclear diffuse scattering begins to decrease at $T \sim 190$ K ($=T_{MD}$) as reported in the previous study [5]. In the low temperature region the fluctuation of PNR is suppressed. The polar directions of PNRs tend to be aligned and the correlation length of the crystal structure is longer. The decrease of the nuclear diffuse scattering intensity below $T_{MD}$ is unusual because the nuclear diffuse scattering in the typical relaxor has no change at low temperature.

At $T_{MD}$, furthermore, the magnetic intensity at $(1/3,1/3,L)$ appears with decreasing temperature, and the magnetic correlation length in the $c$-plane coincides with the nuclear correlation length in the $c$-plane. These results mean the existence of a strong coupling between PNR and magnetic ordering. The alignment of the polar direction in PNR and the growth of the magnetic ordering support each other because the polar crystal structure is expected to release the geometrical frustration.

In conclusion, we have carried out the measurements of the dielectric permittivity and the nuclear diffuse scattering on the single crystal of relaxor magnet $LuFeCoO_4$. The temperature dependence of the dielectric permittivity has the large anisotropy. The fluctuation of the polar direction in polar nanoregion in the $c$-plane is larger than that along the $c$-axis.

Work at JRR-3 was supported by ISSP of the University of Tokyo. This work was supported by KAKENHI (19052002, 20740171, 24740224, and 15K05123).

## References

[1] W. Eerenstein, N. D. Mathur, and J. F. Scott, Nature **442** (2006) 759.
[2] T. Kimura, T. Goto, H. Shintani, K. Ishizaka, T. Arima, and Y. Tokura, Nature **426** (2003) 55.
[3] M. Soda, M. Matsuura, Y. Wakabayashi, and K. Hirota, J. Phys. Soc. Jpn. **80**, (2011) 043705.
[4] A. A. Bokov and Z.-G. Ye, J. Mater. Sci. **41**, (2006) 31.
[5] M. Soda and T. Masuda, J. Phys. Soc. Jpn. **85**, (2016) 034713.
[6] K. Yoshii, N. Ikeda, Y. Matsuo, Y. Horibe, S. Mori, Phys. Rev. B **76**, (2007) 024423.
[7] M. Isobe, N. Kimizuka, and S. Takekawa, Acta Cryst. C **46**, (1990) 1917.
[8] G. Xu, G. Shirane, J. R. D. Copley, and P. M. Gehring, Phys. Rev. B **69**, (2004) 064112.

IOP Conf. Series: Journal of Physics: Conf. Series **828** (2017) 012002 doi:10.1088/1742-6596/828/1/012002

# Effects of magnetic field and hydrostatic pressure on the distorted triangular lattice antiferromagnet $RbFeBr_3$

**Nobuyuki Kurita, Hidekazu Tanaka**

Department of Physics, Tokyo Institute of Technology, Meguro-ku, Tokyo 152-8551, Japan

E-mail: `kurita@phys.titech.ac.jp`

**Abstract.** We report on magnetization measurements of $RbFeBr_3$ single crystals. At ambient pressure, the field–temperature phase diagram has been determined up to 70 kOe for $H \parallel c$. It is found that the application of magnetic field for $H \parallel c$ expands the low temperature phase with the triangular spin structure ($T < T_{N2}$) and suppresses the partially disordered phase ($T_{N2} < T < T_{N1}$). When hydrostatic pressure is applied under 100 Oe for $H \parallel c$, the magnetization anomaly related to $T_{N1}$ shifts to lower temperatures and is no more detectable at $P \geq 1.0$ GPa.

## 1. Introduction

Over the past decades, hexagonal $ABX_3$-type compounds has continued to attract considerable interest as a model system of triangular-lattice antiferromagnets (TLA). The spin chains consisting of magnetic $B^{2+}$ ions centered at $BX_6$ octahedra align along the $c$-axis and form triangular lattices in the $c$-plane. Since the exchange coupling between the spin chains are antiferromagnetic (AF), a rich variety of quantum phenomena have been observed owing to the geometric spin frustration characteristic of TLA [1].

As is the case with typical $ABX_3$ series, $RbFeBr_3$ crystallizes in the hexagonal $CsNiCl_3$ structure with the space group $P6_3/mmc$ at room temperature. With decreasing temperature, this compound undergoes successive structural phase transitions at 109 K and 39.5 K to a lower symmetry phase which is found to be ferroelectric along the $c$-axis [2, 3, 4]. The crystal structure of the low temperature phase is of the hexagonal $KNiCl_3$-type with $P6_3cm$ where the triangular lattices are distorted. Accordingly, there exist two kinds of interlayer exchange interactions so that the spin frustration is partially released in the $c$-plane. Several experiments have revealed that two AF phase transitions occur at $T_{N1} = 5.6$ K and $T_{N2} = 2.0$ K in the ferroelectric phase [5, 6, 7]. The successive phase transition in the XY-type disordered triangular antiferromagnets is consistent with theoretical results [8, 9]. Recent neutron diffraction measurements in the absence of applied field have shown that the spin configuration is partially disordered (PD) in the intermediate phase ($T_{N2} < T < T_{N1}$) and triangular with spin directions slightly tilt from 120° to each other at lower temperatures below $T_{N2}$ [7]. In the PD phase, one of three spin chains is uncorrelated and the other two spin chains couple antiferromagnetically.

It is known that magnetic $Fe^{2+}$ ions in an octahedral environment can be described by effective spin $S = 1$ at low temperatures well below $\lambda/k_B \simeq 150$ K where $\lambda$ is the spin-orbit coupling [10]. Because the $FeBr_6$ octahedrons are trigonally elongated, $RbFeBr_3$ has the singlet ground state and the doublet first excited state which are separated by the large easy-plane single-ion anisotropy $D(S^z)^2$ [11, 12]. In $RbFeBr_3$, exchange interactions are predominant over the magnetic anisotropy so that three dimensional magnetic orders occur in the absence of magnetic field.

In this proceedings, we investigate the field and pressure effects on the magnetic phase transitions of $RbFeBr_3$, via magnetization measurements using single crystals.

## 2. Experimental Details

Single crystals of $RbFeBr_3$ were grown via the vertical Bridgman method from a stoichiometric mixture of constituent elements sealed in an evacuated quartz tube. The ingredients were dehydrated in vacuum by heating up to 150°C for three days. The temperature at the center of the furnace was set at 660°C and the crystals were lowered at a rate of 3 mm/hour. We repeated the same procedure after the removal of impurities and imperfect crystals.

The magnetization measurements were carried out down to 1.8 K under magnetic fields of up to 70 kOe using a SQUID magnetometer (MPMS-XL, Quantum Design). High pressure magnetization was measured up to a pressure of 1.2 GPa using a clamped piston cylinder pressure device. Daphne 7373 (Idemitsu Kosan), which remains in the liquid state up to $\sim$ 2 GPa at room temperature [13], was used as a pressure-transmitting medium. The pressure generated in the sample space was calibrated at a low temperature by the change in the superconducting transition temperature $T_c$ of tin under $H = 10$ Oe. The high-pressure magnetization data are corrected for the background contributions.

## 3. Results and Discussions

Figure 1(a) shows temperature dependence of magnetic susceptibility $\chi\,(=M/H)$ of $RbFeBr_3$ under 1 kOe for $H \parallel c$ and $H \perp c$. The obtained $\chi(T)$ data are in good agreement with previous results [11]. $\chi(T)$ for $H \parallel c$ exhibits a broad peak around 15 K followed by a rapid decrease with decreasing temperature, while $\chi(T)$ for $H \perp c$ shows a continuous increase down to the lower temperatures. The anisotropic behaviors in $\chi(T)$ with respect to the field direction are mostly attributable to the large easy-plane single-ion anisotropy $D(S^z)^2$ due to the trigonal crystalline field [14]. Figure 1(b) shows the low temperature $\chi(T)$ data of $RbFeBr_3$ under 1 kOe. Arrows denote AF transition temperatures $T_{N1} = 5.6$ K and $T_{N2} = 2.0$ K reported by the previous specific heat study in zero field [5]. For $H \parallel c$, one can see two slight anomalies both at $T_{N1}$ and $T_{N2}$. For $H \perp c$, $\chi(T)$ exhibits a clear jump at $T_{N1}$ while no sign associated with $T_{N2}$ was identified.

Below, we focus on results obtained for $H \parallel c$. Figure 2 shows the $\chi(T)$ and $d\chi/dT(T)$ data of $RbFeBr_3$ under several fields up to 70 kOe for $H \parallel c$. As indicated by arrows in the $d\chi/dT(T)$ data under 10 kOe, we assign $T_{N1}$ and $T_{N2}$ as temperatures where $d\chi/dT(T)$ shows a sudden decrease with decreasing temperature. With increasing field, no significant change is found for $T_{N1}$ while the magnetization anomaly related to $T_{N2}$ shifts to higher temperatures and evolves to a cusplike peak. It is noted that the similar cusplike peaks in $\chi(T)$ are observed at the AF transition temperatures for the field-induced ordered phase of the related compounds $CsFeCl_3$ [15] and $CsFeBr_3$ [16]. Contrary to the case with $RbFeBr_3$, these compounds have the gapped singlet ground state in zero field since exchange interactions are not strong enough to produce the long range order [18, 19, 20]. When magnetic field is applied above the critical values $H_c \sim 40$ kOe for $CsFeCl_3$ [15, 17] and $\sim 30$ kOe for $CsFeBr_3$ [16], these compounds exhibit field-induced AF phase transitions. Figure 3 illustrates the magnetic phase diagram of $RbFeBr_3$ up to 70 kOe for $H \parallel c$, determined via the present magnetization measurements. It is found that application of magnetic field for $H \parallel c$ on this compound stabilizes the triangular spin phase (phase II) and suppresses the PD phase (phase I) whereas there is little effect on $T_{N1}$.

Next, we present results of the high-pressure magnetization measurement performed under low magnetic field. Figure 4 shows $\chi(T)$ data of $RbFeBr_3$ at several pressures up to 1.2 GPa

IOP Conf. Series: Journal of Physics: Conf. Series **828** (2017) 012002 doi:10.1088/1742-6596/828/1/012002

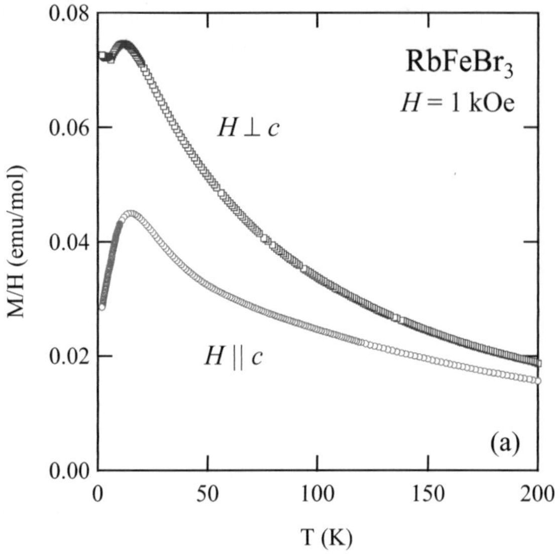


**Figure 1.** (Color online) (a) Magnetic susceptibility $\chi$ (=$M/H$) vs $T$ of $RbFeBr_3$ under 1 kOe for $H \parallel c$ and $H \perp c$. (b) Expanded view of $\chi(T)$ data at low temperatures. $\chi(T)$ for $H \perp c$ is shifted in the longitudinal direction by $-0.03$ emu/mol. Arrows denote $T_{N1}$ and $T_{N2}$ reported by the previous specific heat study [5].

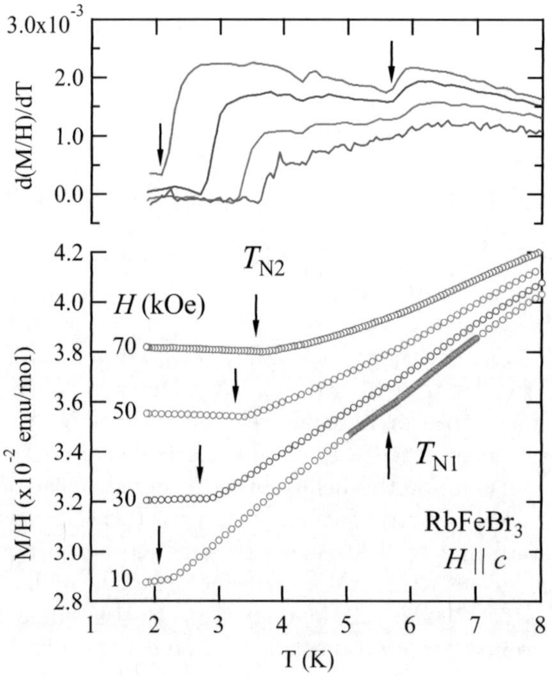


**Figure 2.** (Color online) (a) Temperature dependence of $\chi$ (=$M/H$) and $d\chi/dT$ of $RbFeBr_3$ for $H \parallel c$ under several fields up to 70 kOe. Arrows indicate $T_{N1}$ and $T_{N2}$ deduced from anomalies in $d\chi/dT(T)$ data.

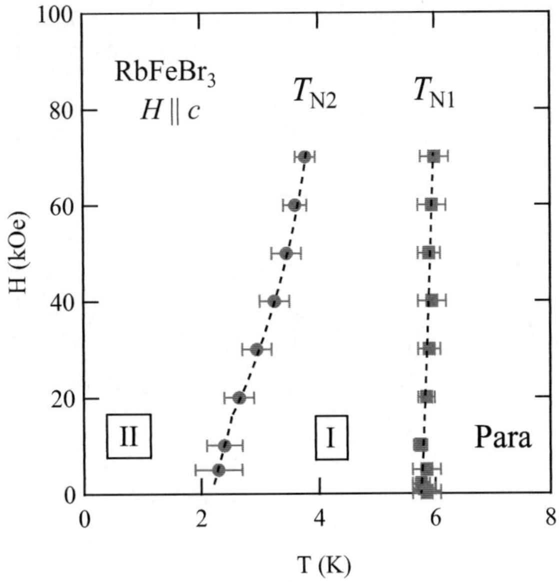


**Figure 3.** (Color online) Field-temperature phase diagram of $RbFeBr_3$ for $H \parallel c$ obtained from the magnetization measurements. Dashed curves are guides to the eyes. Spin configurations in the phases I and II are partially disordered and triangular structures, respectively [5].

under 100 Oe for $H \parallel c$. For clarity, the data are shifted arbitrarily in the longitudinal direction. The lowest measured temperature at high-pressures is limited to 2 − 3 K because of the Meissner effect induced by a superconducting transition of tin, which is included in the sample space as a pressure sensor. At ambient pressure, as compared to results under 1 kOe shown in Fig. 1(b), $\chi(T)$ under 100 Oe exhibits a clear jump at $T_{N1}$ while there is no obvious signature of $T_{N2}$. With increasing pressure, the magnetization anomaly due to $T_{N1}$ shifts to lower temperatures and is no more detectable at $P \geq 1.0$ GPa.

Here, the role of hydrostatic pressure on $RbFeBr_3$ is discussed. Since the intrachain exchange interaction $J_0$ is an order of magnitude larger than the interchain exchange interactions [11, 12], we only consider pressure evolution of $J_0$. In this compound, there exist both AF and ferromagnetic (FM) paths of exchange interactions. Considering the fact that $J_0$ is AF, the AF exchange interaction dominates over the ferromagnetic (FM) exchange interaction. Thus, the primary effect of pressure could be to enhance the FM intrachain exchange interaction since $T_{N1}$ is suppressed by applying hydrostatic pressure. This scenario is consistent with high pressure studies of $CsFeBr_3$ [21] and $CsFeCl_3$ [15]. In $CsFeBr_3$ where $J_0$ is AF, application of hydrostatic pressure decreases the AF transition temperature for the field-induced ordered phase and increases the transition field $H_c$. In $CsFeCl_3$ where $J_0$ is FM, in contrast, the field-induced ordered phase moves toward a higher-temperature and lower-field with increasing hydrostatic pressure, and, eventually, a pressure-induced magnetic phase transition occurs at the critical pressure of 0.9 GPa.

## 4. Conclusions

To conclude, we have performed magnetization measurements of $RbFeBr_3$ single crystals under fields up to 70 kOe at ambient pressure and under 100 Oe at high pressures up to 1.2 GPa. At ambient pressure, the obtained field–temperature phase diagram for $H \parallel c$ indicates that, with increasing magnetic field, the triangular ground state and PD phase become stabilized and

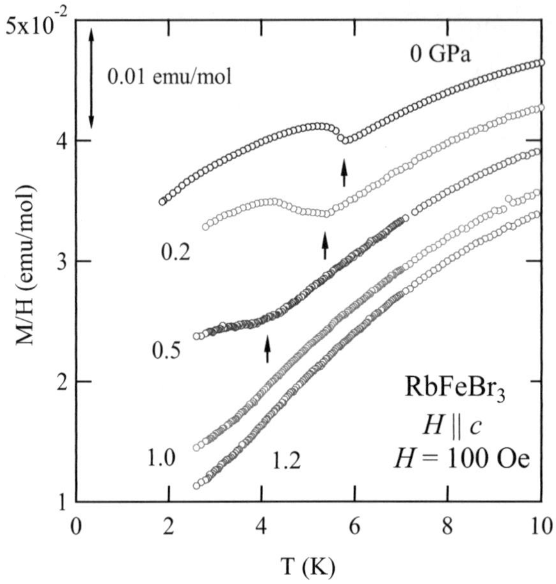


**Figure 4.** (Color online) $M/H$ vs $T$ of $RbFeBr_3$ under 100 Oe for $H \parallel c$ at pressures up to 1.2 GPa. Arrows indicate $T_{N1}$. For clarity, the data are shifted arbitrarily in the longitudinal direction.

suppressed, respectively. With increasing pressure under 100 Oe, $T_{N1}$ decreases and no sign of $T_{N1}$ is detectable at $P \geq 1.0$ GPa. This suggests that the primary role of hydrostatic pressure on this compound could be to enhance the FM intrachain exchange interactions.

**Acknowledgments**
This work was supported by Grants-in-Aid for Scientific Research (A) Nos. 23244072 and 26247058, (C) No. 16K05414 from Japan Society for the Promotion of Science.

## References

[1] Collins M F and Petrenko O A 1997 *Can. J. Phys.* **75**, 605
[2] Eibschütz M, Davidson G R and Cox D. E 1973 *AIP Conf. Proc.* **18**, 386
[3] Harrison A and Visser D 1989 *J. Phys.: Condens. Matter.* **1** 733.
[4] Mitsui T, Machida K, Kato T and Iio K, 1994 *J. Phys. Soc. Jpn.* **63** 839
[5] Achiwa N 1969 *J. Phys. Soc. Jpn.* **27** 561
[6] Kato T, Nishino T, Morishita K, Tanaka H and Iio K 1998 *J. Mag. Mag. Mater.* **177-181** 831
[7] Nishiwaki Y, Oosawa A, Kakurai K, Kaneko K, Tokunaga M and Kato T 2011 *J. Phys. Soc. Jpn.* **80** 084711
[8] Kawamura H 1990 *J. Phys. Soc. Jpn.* **101** 545
[9] Plumer M L, Caillé A and Kawamura H 1991 *Phys. Rev. B* **44** 4461
[10] Inomata K and Oguchi T 1967 *J. Phys. Soc. Jpn.* **23** 771.
[11] Lines M E and Eibschütz M 1975 *Phys. Rev. B* **11** 4583
[12] Harrison A and Visser D 1989 *Phys. Lett.* **137A** 79.
[13] Murata K, Yoshino H, Yadav H O, Honda Y and Shirakawa N 1977 *Rev. Sci. Instrum.* **68**, 2490
[14] Adachi K, Takeda K, Matsubara F, Mekata M and Haseda T 1983 *J. Phys. Soc. Jpn.* **52** 2202
[15] Kurita N and Tanaka H 2016 *Phys. Rev. B* **94** 104409
[16] Tanaka Y, Tanaka H, Ono T, Oosawa A, Morishita K, Iio K, Kato T, Aruga-Katori H, Bartashevich M. I and Goto T 2001 *J. Phys. Soc. Jpn.* **70**, 3068; 2007 *J. Phys. Soc. Jpn.* **76**, 108001
[17] Haseda T, Wada N, Hata M and Amaya K 1981 *Physica B* **108** 841
[18] Yoshizawa H, Kozukue W and Hirakawa K., 1980 *J. Phys. Soc. Jpn.* **49** 144
[19] Bocquet S 1988 *Phys. Rev. B* **37** 7840

[20] Dorner B, Visser D, Steigenberger U, Kakurai K and Steiner M 1988 *Z. Phys. B* **72** 487
[21] Momosaki Y, Ono T, and Tanaka H unpublished data

# Frustrated $S = 1/2$ Two-Leg Ladder with Different Leg Interactions

**Takashi Tonegawa**[1,2], **Kiyomi Okamoto**[3], **Toshiya Hikihara**[4] **and Tôru Sakai**[5,6]

[1]Professor Emeritus, Kobe University, Kobe 657-8501, Japan
[2]Department of Physical Science, Osaka Prefecture University, Sakai, 599-8531, Japan
[3]College of Engineering, Shibaura Institute of Technology, Saitama, 337-8570, Japan
[4]Faculty of Science and Technology, Gunma University, Kiryu, 376-8515, Japan
[5]Graduate School of Material Science, University of Hyogo, Hyogo 678-1297, Japan
[6]National Institutes for Quantum and Radiological Science and Technology (QST), SPring-8, Hyogo 679-5148, Japan

E-mail: tone0115@vivid.ocn.ne.jp

**Abstract.** We explore the ground-state phase diagram of the $S = 1/2$ two-leg ladder. The isotropic leg interactions $J_{l,a}$ and $J_{l,b}$ between nearest neighbor spins in the legs $a$ and $b$, respectively, are different from each other. The $xy$ and $z$ components of the uniform rung interactions are denoted by $J_r$ and $\Delta J_r$, respectively, where $\Delta$ is the $XXZ$ anisotropy parameter. This system has a frustration when $J_{l,a} J_{l,b} < 0$ irrespective of the sign of $J_r$. The phase diagrams on the $\Delta$ $(0 \leq \Delta < 1)$ versus $J_{l,b}$ plane in the cases of $J_{l,a} = -0.2$ and $J_{l,a} = 0.2$ with $J_r = -1$ are determined numerically. We employ the physical consideration, the level spectroscopy analysis of the results obtained by the exact diagonalization method and also the density-matrix renormalization-group method. It is found that the non-collinear ferrimagnetic (NCFR) state appears as the ground state in the frustrated region of the parameters. Furthermore, the direct-product triplet-dimer (TD) state in which all rungs form the TD pair is the *exact ground state*, when $J_{l,a} + J_{l,b} = 0$ and $0 \leq \Delta \lesssim 0.83$. The obtained phase diagrams consist of the TD, $XY$ and Haldane phases as well as the NCFR phase.

## 1. Introduction

In the past years a great deal of work has been devoted to the study which aims at clarifying the role of the frustration in low-dimensional quantum spin systems with competing interactions. As regards the $S = 1/2$ two-leg ladder systems, the general cases where additional leg next-nearest-neighbor and/or diagonal interactions are competing with the leg nearest-neighbor and rung interactions have been extensively investigated [1–3]. Very recently, we [4] have discussed the ground-state phase diagram of the frustrated $S = 1/2$ two-leg ladder, in which rung interactions are ferromagnetically-antiferromagnetically alternating and have a common Ising-type anisotropy, while leg interactions are antiferromagnetically uniform and isotropic. The phase diagram which we have numerically determined in the case where the leg interactions are relatively weak compared with the rung interactions shows that the incommensurate Haldane state as well as the commensurate one appears as the ground state in the whole range of the Ising-type anisotropy parameter. This appearance of the Haldane state in the case where the Ising character of rung interactions is strong is contrary to the ordinary situation, and is called the inversion phenomenon concerning the interaction anisotropy [5–8]. The ground-state phase diagram of the frustrated rung-alternating $S=1/2$ two-leg ladder in which all interactions are isotropic has also been studied by combining analytical approaches with numerical simulations [9]. Furthermore, it has been shown that the introduction of the rung alternation gives rise to the half-magnetization plateau in the ground-sate magnetization curve [10]. This result is consistent with the necessary condition for the appearance of the magnetization plateau by Oshikawa, Yamanaka and Affleck [11].

In the present paper, we explore the ground-state phase diagram of another frustrated $S=1/2$ two-leg ladder with different leg interactions. We express the Hamiltonian which describes this system as

$$\mathcal{H} = J_{1,a}\sum_{j=1}^{L}\vec{S}_{j,a}\cdot\vec{S}_{j+1,a} + J_{1,b}\sum_{j=1}^{L}\vec{S}_{j,b}\cdot\vec{S}_{j+1,b} + J_{\rm r}\sum_{j=1}^{L}\{S^x_{j,a}S^x_{j,b}+S^y_{j,a}S^y_{j,b}+\Delta S^z_{j,a}S^z_{j,b}\}\,. \quad (1)$$

Here, $\vec{S}_{j,\ell}=(S^x_{j,\ell}, S^y_{j,\ell}, S^z_{j,\ell})$ is the $S=1/2$ operator acting at the $(j,\ell)$ site assigned by rung $j$ and leg $\ell(=a$ or $b)$; $J_{1,a}$ and $J_{1,b}$ denote, respectively, the magnitudes of the isotropic leg $a$ and leg $b$ interactions; $J_{\rm r}$ denotes that of the anisotropic rung interaction, the $XXZ$-type anisotropy being controlled by the parameter $\Delta$; $L$ is the total number of rungs, which is assumed to be even. The sketch of the present model is given in Fig. 1. It should be noted that this system has a frustration when $J_{1,a}J_{1,b}<0$ irrespective of the sign of $J_{\rm r}$.

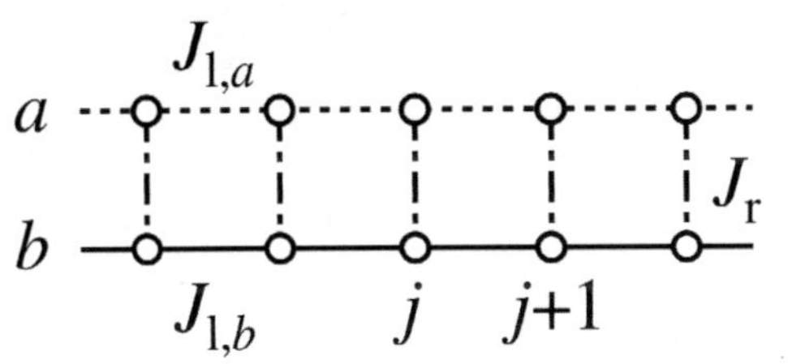


**Figure 1.** Sketch of the present model. Open circles denote $S=1/2$ spins, and lines three kinds of interactions between spins.

The most characteristic feature of the present system is the fact that, when the condition $J_{1,a}+J_{1,b}=0$, which belongs to the frustration region, is satisfied, the following three states are the *exact eigenstates* of the Hamiltonian (1).

1) The direct-product singlet-dimer (SD) state in which all rungs form the SD $((\alpha_{j,a}\beta_{j,b}-\beta_{j,a}\alpha_{j,b})/\sqrt{2})$ pair.
2) The direct-product triplet-dimer (TD) state in which all rungs form the TD $((\alpha_{j,a}\beta_{j,b}+\beta_{j,a}\alpha_{j,b})/\sqrt{2})$ pair.
3) The nematic state with an arbitrary phase $\phi$ in which all rungs are in the state given by a linear combination of two ferromagnetic states, $\cos\phi\,\alpha_{j,a}\alpha_{j,b}+\sin\phi\,\beta_{j,a}\beta_{j,b}$.

Here, $\alpha_{j,l}$ denotes the $S^z_{j,l}=+1/2$ state and $\beta_{j,l}$ the $S^z_{j,l}=-1/2$ state. These facts can be proven by operating the Hamiltonian (1) directly to the above three states. Furthermore, it can be analytically shown that, when $J_{1,a}+J_{1,b}=0$, $J_{\rm r}<0$, and the $XY$-type anisotropy of rung interactions is sufficiently strong $\left(1-\frac{2|J_{1,a}|}{|J_{\rm r}|}\gg\Delta\geq 0\right)$, the direct-product TD state is the *exact ground state* of the system, and that, when $J_{1,a}+J_{1,b}=0$ and $J_{\rm r}(>0)$ is sufficiently large, the direct-product SD state is the *exact ground state* of the system. It is noted that the above results concerning with the direct-product SD state has already been shown by Tsukano and Takahshi [12]. We also note that all of the above results including the nematic state with $\phi$ as well as the direct-product TD and SD states are applicable to systems in higher dimensions, in which units of two $S=1/2$ spins form lattices; the details will be discussed in our forthcoming paper [13].

Unfortunately, materials corresponding to the present model have been neither yet found nor synthesized so far. We believe, however, that it is a physically realistic model. In fact, for example, Yamaguchi *et al.* [14, 15] have recently demonstrated the modulation of magnetic interactions in spin ladder systems by using verdazyl-radical crystals. It is highly expected that the flexibility of molecular arrangements in such organic-radical materials realizes $S=1/2$ two-leg ladder systems with different leg interactions.

In the following discussions, we confine ourselves to the case where $J_{\rm r}$ is ferromagnetic, and we put $J_{\rm r}=-1$, choosing $|J_{\rm r}|$ as the unit of energy. Then, when $0<|J_{{\rm l},\ell}|\ll 1$, the present ladder system can be mapped onto the $S=1$ chain by using the degenerate perturbation theory. We discuss this mapping in the next section (section 2). Section 3 is devoted to the discussions on the ground-state phase diagram. Assuming, for simplicity, that $J_{{\rm l},a}=-0.2$ or $0.2$ and $0\le\Delta<1$ (the $XY$-type anisotropy of rung interactions), we determine the ground-state phase diagrams on the $\Delta$ versus $J_{{\rm l},b}$ plane. We mainly use the numerical methods such as the exact-diagonalization (ED) method and the density-matrix renormalization-group (DMRG) method [16, 17] with the help of physical considerations. Finally, we give concluding remarks in section 4.

## 2. Mapping onto the $S=1$ chain

We discuss the case where $0<|J_{{\rm l},\ell}|\ll|J_{\rm r}|$, assuming that $J_{\rm r}=-1$. The four eigenstates for rung $j$ are given by $\psi_j^{(1,+)}=\alpha_{j,a}\alpha_{j,b}$, $\psi_j^{(1,0)}=(\alpha_{j,a}\beta_{j,b}+\beta_{j,a}\alpha_{j,b})/\sqrt{2}$, $\psi_j^{(1,-)}=\beta_{j,a}\beta_{j,b}$ and $\psi_j^{(0,0)}=(\alpha_{j,a}\beta_{j,b}-\beta_{j,a}\alpha_{j,b})/\sqrt{2}$, and the corresponding energies are, respectively, $E^{(1,+)}=-\Delta/4$, $E^{(1,0)}=(\Delta-2)/4$, $E^{(1,-)}=-\Delta/4$ and $E^{(0,0)}=(\Delta+2)/4$, for all $j$'s. Thus, the state $\psi_j^{(0,0)}$ can be neglected. We introduce the pseudo $S=1$ operator $\vec{T}_j$ for rung $j$, and make the $T_j^z=+1$, $0$ and $-1$ states correspond to the $\psi_j^{(1,+)}$, $\psi_j^{(1,0)}$ and $\psi_j^{(1,-)}$ states, respectively. The relation $\vec{T}_j=\vec{S}_{j,a}+\vec{S}_{j,b}$ holds, as is readily shown by comparing the matrix elements of both operators $\vec{T}_j$ and $\vec{S}_{j,l}$ in the subspace of $\phi_j^{(1,+)}$, $\phi_j^{(1,0)}$ and $\phi_j^{(1,-)}$. Thus, the Hamiltonian (1) for the $S=1/2$ operator $\vec{S}_{j,l}$ can be mapped onto the effective Hamiltonian $\mathcal{H}_{\rm eff}$ for the $S=1$ operator $\vec{T}_j$, which is given by

$$\mathcal{H}_{\rm eff}=J_{\rm eff}\sum_{j=1}^{L}\vec{T}_j\cdot\vec{T}_{j+1}+D_{\rm eff}\sum_{j=1}^{L}(T_j^z)^2;\quad J_{\rm eff}=\frac{J_{{\rm l},a}+J_{{\rm l},b}}{4},\quad D_{\rm eff}=\frac{(1-\Delta)}{2},\tag{2}$$

where $T_j^z$ is the $z$-component of $\vec{T}_j$. It is noted that the on-site anisotropy ($D_{\rm eff}$-) term comes from the difference between $E^{(1,+)}=E^{(1,-)}$ and $E^{(1,0)}$.

The above $\mathcal{H}_{\rm eff}$ is the result of the degenerate perturbation calculation in the lowest-order of $|J_{{\rm l},\ell}|/|J_{\rm r}|$. It is apparent that this is not applicable to discussing the frustrated region of the original Hamiltonian (1), which includes the case of $J_{{\rm l},a}+J_{{\rm l},b}=0$. In order to improve this point, higher-order perturbation calculations are indispensable; these calculations are left for a future study.

The ground-state phase diagram of the anisotropic $S=1$ chain has been determined by several authors [18–20]. According to their results, as the value of $D_{\rm eff}$ increases from zero, the phase transition from the $XY$ (or Haldane) phase to the large-$D$ phase takes place at $D_{\rm eff}\simeq 1$ when $J_{\rm eff}=-1$ (or when $J_{\rm eff}=1$). Thus, we may expect that in our $S=1/2$ ladder with $J_{\rm r}=-1$, the phase transition between the $XY$ and TD phases occurs at $J_{{\rm l},a}+J_{{\rm l},b}\simeq 2(\Delta-1)$ when $J_{{\rm l},a}+J_{{\rm l},b}<0$ (or, equivalently, when $\Delta<1$), and also that the phase transition between the Haldane and TD phases occurs at $J_{{\rm l},a}+J_{{\rm l},b}\simeq 2(1-\Delta)$ when $J_{{\rm l},a}+J_{{\rm l},b}>0$ (or, when $\Delta<1$, again). It is noted that the large-$D$ state in the spin-1 chain is equivalent to the TD state in the present $S=1/2$ ladder, since in the valence bond picture of the former state, each $S=1$ spin consists of two $S=1/2$ spins forming the TD pair, as is well known.

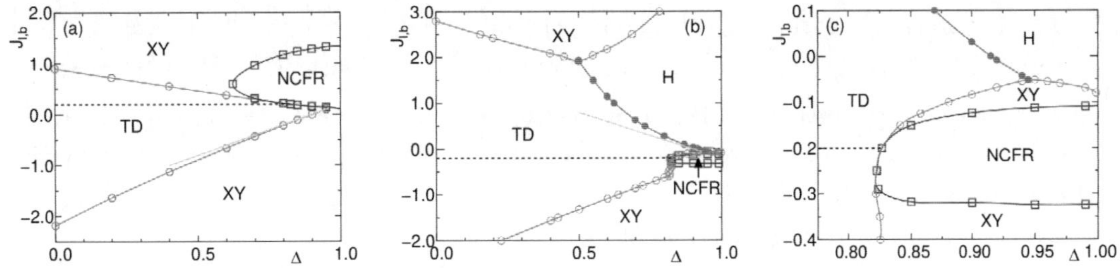


**Figure 2.** Ground-state phase diagram on the $\Delta$ versus $J_{1,b}$ plane for (a) $J_{1,a}=-0.2$ and (b,c) $J_{1,a}=0.2$ with $J_{\rm r}=-1$; in (c) part of (b) is enlarged. The regions designated by TD, XY, NCFR and H are, respectively, those of the triplet-dimer, $XY$, non-collinear ferrimagnetic and Haldane phases. See the text for the meanings of several lines.

## 3. Ground-state phase diagrams

Throughout this section we assume that $J_{\rm r}=-1$, as mentioned before. Figure 2 shows the ground-state phase diagrams on the $\Delta$ versus $J_{1,b}$ plane determined for $J_{1,a}=-0.2$ and $J_{1,a}=0.2$. The former phase diagram consists of the TD, $XY$ and non-collinear ferrimagnetic (NCFR) phases [12, 21], and in the latter one, the Haldane (H) phase appears in addition to the above three phases. There are three kinds of the phase transition lines, which we have numerically estimated as discussed below in detail. The magenta lines with open circles are the phase transition lines between the TD or H phase and the $XY$ phase which are of the Berezinskii-Kosterlitz-Thouless (BKT) type [22, 23], the red line with closed circles is the phase transition line between the TD and H phases which are of the Gaussian-type, and finally the blue lines with open squares are the phase transition lines between the NCFR phase and the TD or $XY$ phase. In the latter phase diagram, there are two tricritical points at $(\Delta, J_{1,b})=(0.500(1), 1.917(1))$ and $(0.945(1), -0.052(1))$ associated with TD, $XY$ and H phases. The green straight lines show the results of the comparison of the degenerate perturbation calculations with the numerical results [18–20] (see section 2); in Fig. 2(a) it is for the TD-$XY$ transition and given by $J_{1,b}=2\Delta-1.8$, while in Fig. 2(b) it is for the TD-H transition and given by $J_{1,b}=-2\Delta+1.8$. In both cases they are in excellent agreement with the numerical results at least when $|J_{1,b}|$ is not too large. It is noted that on the special lines where $J_{1,a}+J_{1,b}=0$, which are shown by the black broken lines, the direct-product TD state is the *exact ground state*.

In the following explanations for the estimation of the above phase boundary lines, we denote, respectively, by $E_0(L,M;{\rm pbc})$ and $E_1(L,M;{\rm pbc})$ the lowest and second-lowest energy eigenvalues of the Hamiltonian (1) within the subspace determined by $L$ and $M$ under periodic boundary conditions, $\vec{S}_{L+1,\ell}=\vec{S}_{1,\ell}$. The quantity $M$ is the total magnetization given by $M=\sum_{j=1}^{L}(S_{j,a}^z+S_{j,b}^z)$, which is a good quantum number with the eigenvalues of $M=0$, $\pm1$, $\cdots$, $\pm L$. Similarly, we also denote by $E_0(L,M,P;{\rm tbc})$ the lowest energy eigenvalue of the Hamiltonian (1) within the subspace determined by $L$, $M$ and $P$ under twisted boundary conditions, $S_{L+1,\ell}^x=-S_{1,\ell}^x$, $S_{L+1,\ell}^y=-S_{1,\ell}^y$ and $S_{L+1,\ell}^z=S_{1,\ell}^z$, where $P(=+1$ or $-1)$ is the eigenvalue of the space inversion operator with respect to the twisted bond, $\vec{S}_{j,\ell}\leftrightarrow\vec{S}_{L+1-j,\ell}$. We further denote by $E_0(L,M;{\rm obc})$ the lowest energy eigenvalue of the Hamiltonian (1) within the subspace determined by $L$ and $M$ under open boundary conditions, where the sums over $j$ for leg interactions are taken from $j=1$ to $L-1$.

The most powerful method to estimate numerically the phase boundary lines between two of the TD, $XY$ and H phases is the level spectroscopy (LS) method developed by Okamoto, Nomura and Kitazawa [24–27]. In this method, the following three excitation energies [28], $\Delta E_{02}^{(\rm p)}(L)=E_0(L,2;{\rm pbc})-E_0(L,0;{\rm pbc})$, $\Delta E_{00}^{(\rm p,t)}(L,+1)=E_0(L,0,+1;{\rm tbc})-E_0(L,0;{\rm pbc})$ and

$\Delta E_{00}^{(\mathrm{p,t})}(L,-1)=E_0(L,0,-1;\mathrm{tbc})-E_0(L,0;\mathrm{pbc})$ should be compared in the thermodynamic ($L\to\infty$) limit. More strictly speaking, the critical value $J_{1,b\,(\mathrm{cr})}^{(XY,\mathrm{TD})}$ of the BKT $XY$-TD transition, the critical value $J_{1,b\,(\mathrm{cr})}^{(XY,\mathrm{H})}$ of the BKT $XY$-H transition and the critical value $J_{1,b\,(\mathrm{cr})}^{(\mathrm{TD,H})}$ of the Gaussian TD-H transition, which are all for given values of $J_{1,a}$ and $\Delta$, are estimated as follows. First, the corresponding finite-size critical values $J_{1,b\,(\mathrm{cr})}^{(XY,\mathrm{TD})}(L)$, $J_{1,b\,(\mathrm{cr})}^{(XY,\mathrm{H})}(L)$ and $J_{1,b\,(\mathrm{cr})}^{(\mathrm{TD,H})}(L)$ are estimated, respectively, by solving numerically the equations [29],

$$\Delta E_{02}^{(\mathrm{p})}(L) = \Delta E_{00}^{(\mathrm{p,t})}(L,+1) < \Delta E_{00}^{(\mathrm{p,t})}(L,-1)\,, \tag{3}$$

$$\Delta E_{02}^{(\mathrm{p})}(L) = \Delta E_{00}^{(\mathrm{p,t})}(L,-1) < \Delta E_{00}^{(\mathrm{p,t})}(L,+1)\,, \tag{4}$$

$$\Delta E_{00}^{(\mathrm{p,t})}(L,+1) = \Delta E_{00}^{(\mathrm{p,t})}(L,-1) < \Delta E_{02}^{(\mathrm{p})}(L)\,. \tag{5}$$

Then, these finite-size results are extrapolated to the $L\to\infty$ limit to obtain, respectively, the critical values, $J_{1,b\,(\mathrm{cr})}^{(XY,\mathrm{TD})}$, $J_{1,b\,(\mathrm{cr})}^{(XY,\mathrm{H})}$ and $J_{1,b\,(\mathrm{cr})}^{(\mathrm{TD,H})}$.

Practically, we have made the ED calculations to estimate $J_{1,b\,(\mathrm{cr})}^{(XY,\mathrm{TD})}(L)$, $J_{1,b\,(\mathrm{cr})}^{(XY,\mathrm{H})}(L)$ and $J_{1,b\,(\mathrm{cr})}^{(\mathrm{TD,H})}(L)$ for finite-$L$ systems with $2L=12,\ 16,\ \cdots,\ 28$ spins. The procedures for these estimations are shown in Fig. 3, for example, for $L=14$, $J_{1,a}=0.2$ and $\Delta=0.90$ or $0.97$. Performing the $L\to\infty$ extrapolations of the above finite-size critical values, we have fitted them to quadratic functions of $1/L^2$ by use of the least-square method, as explained in Fig. 4, for example, for $J_{1,a}=0.2$ and $\Delta=0.90$ or $0.97$, again. Then, as the results of the extrapolations, we have obtained in the $J_{1,a}=0.2$ case, $J_{1,b\,(\mathrm{cr})}^{(XY,\mathrm{TD})}=-0.0833(1)$ and $J_{1,b\,(\mathrm{cr})}^{(\mathrm{TD,H})}=0.0309(1)$ for $\Delta=0.90$, and also $J_{1,b\,(\mathrm{cr})}^{(XY,\mathrm{H})}=-0.0593(1)$ for $\Delta=0.97$. The phase transition lines shown by the magenta and red lines in Fig. 2(b) and (c) are drawn by plotting, as functions of $\Delta$, the values of $J_{1,b\,(\mathrm{cr})}^{(XY,\mathrm{TD})}$, $J_{1,b\,(\mathrm{cr})}^{(\mathrm{TD,H})}$, and $J_{1,b\,(\mathrm{cr})}^{(XY,\mathrm{H})}$ calculated for various values of $\Delta$. Similarly, the phase transition lines shown by the magenta lines in Fig. 2(a) are obtained by calculating $J_{1,b\,(\mathrm{cr})}^{(XY,\mathrm{TD})}$ and $J_{1,b\,(\mathrm{cr})}^{(\mathrm{TD,H})}$ for various values of $\Delta$ in the case of $J_{1,a}=-0.2$.

Let us denote by $M_{\mathrm{g}}(L;\mathrm{obc})$ the ground-state magnetization for the system with $2L$ spins under open boundary conditions, which is the value of $M$ giving the lowest value of

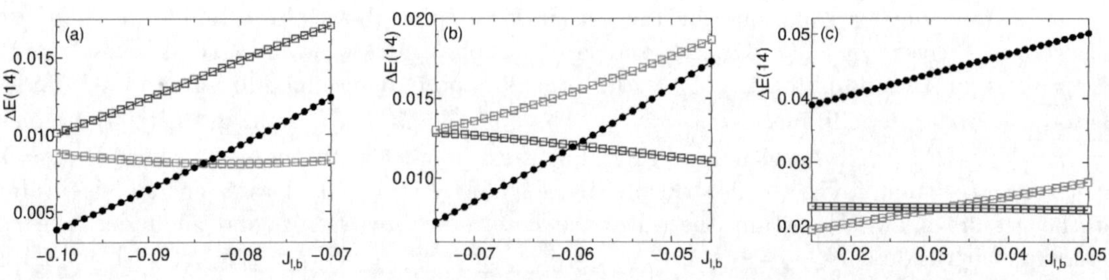


**Figure 3.** Examples of the $\Delta E_{02}^{(\mathrm{p})}(14)$ (black closed circles), $\Delta E_{00}^{(\mathrm{p,t})}(14,+1)$ (magenta open squares) and $\Delta E_{00}^{(\mathrm{p,t})}(14,-1)$ (blue open squares) versus $J_{1,b}$ curves for (a,c) $\Delta=0.90$ and (b) $\Delta=0.97$ with $J_{1,a}=0.2$ and $J_{\mathrm{r}}=-1$ (see Fig. 2(c).) In (a), (b) and (c) we obtain, respectively, $J_{1,b\,(\mathrm{cr})}^{(XY,\mathrm{TD})}(14)=-0.084049$ from the crossing point of the black and magenta curves, $J_{1,b\,(\mathrm{cr})}^{(XY,\mathrm{H})}(14)=-0.059748$ from the crossing point of the black and blue curves and $J_{1,b\,(\mathrm{cr})}^{(\mathrm{TD,H})}(14)=0.031604$ from the crossing point of the magenta and blue curves.

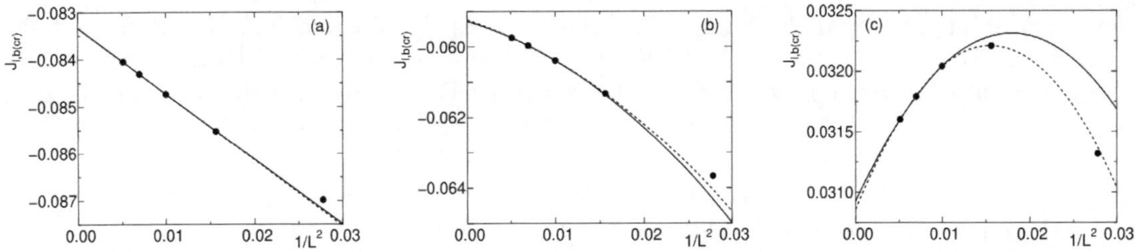


**Figure 4.** Examples of the $L\to\infty$ extrapolations of (a) $J_{1,b\,(\mathrm{cr})}^{(XY,\mathrm{TD})}(L)$ for $\Delta=0.90$, (b) $J_{1,b\,(\mathrm{cr})}^{(XY,\mathrm{H})}(L)$ for $\Delta=0.97$ and (c) $J_{1,b\,(\mathrm{cr})}^{(\mathrm{TD},\mathrm{H})}(L)$ for $\Delta=0.90$, where $J_{1,a}=0.2$ and $J_{\mathrm{r}}=-1$ (see Fig. 2(c).) We assume that these finite-size critical values are quadratic functions of $1/L^2$. The broken lines represent the least-square fittings by use of $L=14$, 12, 10 and 8 data, while the solid lines those without $L=8$ data. From these extrapolations, we obtain $J_{1,b\,(\mathrm{cr})}^{(XY,\mathrm{TD})}=-0.0833(1)$ in (a), $J_{1,b\,(\mathrm{cr})}^{(XY,\mathrm{H})}=-0.0593(1)$ in (b) and $J_{1,b\,(\mathrm{cr})}^{(\mathrm{TD},\mathrm{H})}=0.0309(1)$ in (c), where the numerical errors are estimated from the difference between the extrapolated results with and without the $L=8$ data.

$E_0(L,M;\mathrm{obc})$'s. In the NCFR phase, $M_{\mathrm{g}}(L;\mathrm{obc})$ is finite $(0<M_{\mathrm{g}}(L;\mathrm{obc})<L)$, while in other phases, $M_{\mathrm{g}}(L;\mathrm{obc})=0$ [28]. We have carried out DMRG calculations [16, 17] for the finite system with $2L=72$ spins to estimate the ground-state magnetization per spin, $m_{\mathrm{g}}(L;\mathrm{obc})$, which is defined by $m_{\mathrm{g}}(L;\mathrm{obc})=M_{\mathrm{g}}(L;\mathrm{obc})/(2L)$. The obtained results in the case where $J_{1,a}=-0.2$ and $\Delta=0.8$ are depicted in Fig. 5(a). We see from this figure that the phase transition from the TD phase to the NCFR phase and that from the NCFR phase to the $XY$ phase successively occur with increasing $J_{1,b}$ (see Fig. 2(a)). The finite-size critical values for the former and latter transitions in the $J_{1,a}=-0.2$ and $\Delta=0.8$ case are given, respectively, by $J_{1,b\,(\mathrm{cr})}^{(\mathrm{TD},\mathrm{NCFR})}(36)=0.2185(5)$ and $J_{1,b\,(\mathrm{cr})}^{(\mathrm{NCFR},XY)}(36)=1.1675(5)$. We have performed these DMRG calculations for various $\Delta$'s with $J_{1,a}$ fixed at $J_{1,a}=-0.2$, and obtained the phase transition line shown by the blue line in Fig. 2(a), supposing that the results in the $L=36$ system give good approximate results in the $L\to\infty$ limit [30]. Similarly, the phase transition line shown by the blue line in Fig. 2(b,c) has been obtained by means of the DMRG calculations in the case of

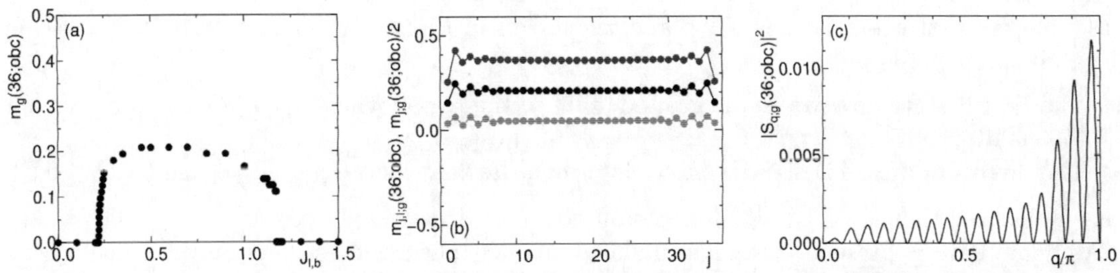


**Figure 5.** (a) Plot of $m_{\mathrm{g}}(36;\mathrm{obc})$ versus $J_{1,b}$, (b) those of $m_{j;\mathrm{g}}(36;\mathrm{obc})/2$ (black line), $m_{j,a;\mathrm{g}}(36;\mathrm{obc})$ (blue line) and $m_{j,b;\mathrm{g}}(36;\mathrm{obc})$ (magenta line) versus $j$ and (c) that of $|S_{q;\mathrm{g}}(36;\mathrm{obc})|^2$ versus $q/\pi$. The quantity $m_{\mathrm{g}}(36;\mathrm{obc})$ in (a) is obtained in the case where $J_{1,a}=-0.2$, $\Delta=0.8$ and $J_{\mathrm{r}}=-1$, while those $m_{j;\mathrm{g}}(36;\mathrm{obc})$ and $|S_{q;\mathrm{g}}(36;\mathrm{obc})|^2$ in (b,c) are obtained in the case where $J_{1,a}=-0.2$, $J_{1,b}=0.7$, $\Delta=0.8$ and $J_{\mathrm{r}}=-1$. (see Fig. 2(a).)

$J_{l,a}=0.2$.

Figure 5(a) suggests that the phase transition between the TD and NCFR phases is of the second order, while that between the NCFR and $XY$ phases is of the first order. However, it is fairly difficult to clarify the order of the phase transition by using only the results of DMRG calculations.

We have also calculated the ground-state site magnetization $m_{j,\ell;\mathrm{g}}(L;\mathrm{obc})$ by use of the DMRG method [16, 17]. This quantity is defined by $m_{j,\ell;\mathrm{g}}(L;\mathrm{obc})=\langle S^z_{j,\ell}\rangle_{L;\mathrm{g}}$, where $\langle\cdots\rangle_{L;\mathrm{g}}$ denotes the expectation value with respect to the ground state of the Hamiltonian (1) under open boundary conditions. Of course, the relation $\sum_{j=1}^{L}\{m_{j,a;\mathrm{g}}(L;\mathrm{obc})+m_{j,b;\mathrm{g}}(L;\mathrm{obc})\}=M_{\mathrm{g}}(L;\mathrm{obc})$ holds. In Fig. 5(b) we plot the $j$-dependences of the ground-state rung magnetization $m_{j;\mathrm{g}}(L;\mathrm{obc})(=m_{j,a;\mathrm{g}}(L;\mathrm{obc})+m_{j,b;\mathrm{g}}(L;\mathrm{obc}))$ and $m_{j,\ell;\mathrm{g}}(L;\mathrm{obc})$, calculated for the $L=36$ system in the case where $J_{l,a}=-0.2$, $J_{l,b}=0.7$ and $\Delta=0.8$; for these parameters $M_{\mathrm{g}}(L;\mathrm{obc})=15$. This figure demonstrates that the $j$-dependences of these quantities are not uniform especially near both of open boundaries. Paying attention to this fact, we have examined the Fourier transform $S_{q;\mathrm{g}}(L;\mathrm{obc})$ of $m_{j;\mathrm{g}}(L;\mathrm{obc})$ [31, 32], defined by

$$S_{q;\mathrm{g}}(L;\mathrm{obc}) = \frac{1}{\sqrt{L}}\sum_{j=1}^{L}\exp(iqj)\left\{m_{j;\mathrm{g}}(L;\mathrm{obc}) - \frac{M_{\mathrm{g}}(L;\mathrm{obc})}{2L}\right\}, \qquad (6)$$

where $q$ is the wave number. The squared modulus $|S_{q;\mathrm{g}}(L;\mathrm{obc})|^2$ of this quantity, calculated for the $L=36$ system in the $J_{l,a}=-0.2$, $J_{l,b}=0.7$ and $\Delta=0.8$ case, where $M_{\mathrm{g}}(L;\mathrm{obc})=15$, is plotted as a function of $q/\pi$ in Fig. 5(c). This figure shows that the largest peak of $|S_{q;\mathrm{g}}(36;\mathrm{obc})|^2$ appears at the position closest to $q=\pi$, suggesting that the wave number of the dominant excitation in the NCFR state is $q=\pi$. (Note that in the system with even $L$ under open boundary conditions, $|S_{q;\mathrm{g}}(L;\mathrm{obc})|^2$ at $q=\pi$ is exactly zero because of the space-inversion symmetry $m_{j;\mathrm{g}}(L;\mathrm{obc}) = m_{L+1-j;\mathrm{g}}(L;\mathrm{obc})$.) We therefore expect that the NCFR state has a commensurate character. In order to examine the commensurability of the NCFR state in full detail, it is necessary to treat the Fourier transform of the rung magnetization $m_{j;\mathrm{g}}(L;\mathrm{obc})$ as well as that of the ground-state two-spin correlation function $\langle S^z_{j,\ell}S^z_{j',\ell}\rangle_{L;\mathrm{g}}$ in larger systems. We will discuss this problem in the near future.

## 4. Concluding remarks

We have numerically determined, with the help of some physical considerations, the ground-state phase diagrams of the $S=1/2$ two-leg ladder with different leg interactions, which is governed by the Hamiltonian (1), in the cases where $J_{l,a}=\pm0.2$, $J_{\mathrm{r}}=-1$ and $0\le\Delta<1$. The obtained phase diagrams on the $\Delta$ versus $J_{l,b}$ plane are shown in Fig. 2. The characteristic features of the results are as follows:

1) The NCFR state appears as the ground state in the region where $J_{l,a}J_{l,b}<0$, when $\Delta$ is not too small.
2) The direct-product TD state is the *exact ground state*, when $J_{l,a}+J_{l,b}=0$ and $0\le\Delta\lesssim0.83$.

It is emphasized that these results are attributed to the frustration effect.

We hope that the present research stimulates future experimental studies on related subjects, which include the synthesization of spin ladder systems with different leg interactions.

## Acknowledgments

We would like to express our sincere thanks to Professors K Hida and H Yamaguchi for their invaluable discussions and comments. This work has been partly supported by JSPS KAKENHI Grant Numbers 15K05198, 16K05419 and 15K05882 (J-Physics) and also by Hyogo Science and Technology Association. Finally, we thank the Supercomputer Center, Institute for Solid State

Physics, University of Tokyo and the Computer Room, Yukawa Institute for Theoretical Physics, Kyoto University for computational facilities.

## References

[1] Lavarélo A, Guillaume G and Laflorencie N 2011 *Phys. Rev.* B **84** 144407 and references therein
[2] Vekua T and Honecker A 2006 *Phys. Rev.* B **73** 214427 and references therein
[3] Michaud F, Coletta T, Manmana S R, Picon J-D and Mila F 2010 *Phys. Rev.* B **81** 014407 and references therein
[4] Tonegawa T, Okamoto K, Hikihara T and Sakai T 2016 *J. Phys.: Conf. Series* **683** 012039
[5] Okamoto K and Ichikawa Y 2002 *J. Phys. Chem. Solids* **63** 1575
[6] Okamoto K 2002 *Prog. Theor. Phys.* Suppl. No.145 208
[7] Tokuno A and Okamoto K 2005 *J. Phys. Soc. Jpn.* **74** Suppl. 157
[8] Okamoto K 2014 *JPS Conf. Proc.* **1** 012031
[9] Amiri F, Sun G, Mikeska H-J and Vekua T 2015 *Phys. Rev.* B **92** 184421
[10] Japaridze G I and Pogosyan E 2006 *J. Phys.: Condens. Matter* **18** 9297
[11] Oshikawa M, Yamanaka M and Affleck I 1997 *Phys. Rev. Lett.* **78** 1984
[12] Tsukano M and Takahashi M 1997 *J. Phys. Soc. Jpn.* **66** 1153
[13] Hikihara T, Tonegawa T, Okamoto K and Sakai T in preparation
[14] Yamaguchi H, Iwase K, Ono T, Shimokawa T, Nakano H, Shimura Y, Kase N, Kittaka S, Sakakibara T, Kawakami T, and Hosokoshi Y 2013 *Phys. Rev. Lett.* **110** 157205
[15] Yamaguchi H, Miyagai H, Shimokawa T, Iwase K, Ono T, Kono Y, Kase N, Araki K, Kittaka S, Sakakibara T, Kawakami T, Okunishi K and Hosokoshi Y 2014 *J. Phys. Soc. Jpn.* **83** 033707
[16] White S R 1992 *Phys. Rev. Lett.* **69** 2863
[17] White S R 1993 *Phys. Rev.* B **48** 10345
[18] Chen W, Hida K and Sanctuary B C 2003 *Phys. Rev.* B **67** 104401
[19] Hu S, Normand B, Wang X and Yu L 2011 *Phys. Rev.* B **84** 220402(R)
[20] Wierschem K and Sengupta P 2014 *Phys. Rev.* B **90** 115157
[21] Yoshikawa S and Miyashita S 2005 *J. Phys. Soc. Jpn. Suppl.* **74** 71
[22] Berezinskii Z L 1971 *Sov. Phys. JETP* **34** 610
[23] Kosterlitz J M and Thouless D J 1973 *J. Phys.* C **6** 1181
[24] Okamoto K and Nomura K 1992 *Phys. Lett.* A **169** 433
[25] Nomura K and Okamoto K 1994 *J. Phys.* A **27** 5773
[26] Kitazawa A 1997 *J. Phys.* A **30** L285
[27] Nomura K and Kitazawa A 1998 *J. Phys.* **31** 7341
[28] We note that in the TD, $XY$ and H phase regions, $E_0(L, 0; \text{pbc})$ always gives the ground-state energy of the finite-$L$ system under periodic boundary conditions.
[29] Here, to solve the equation $A = B < C$ means that to solve the equation $A = B$ under the conditions $A < C$ and $B < C$ (see Fig. 3).
[30] We have made sure that, for a few values of $J_{1,a}$ and $\Delta$, the results in the $L = 48$ system are not so much different from those in the $L = 36$ system.
[31] Hikihara T, Kecke L, Momoi T and Furusaki A 2008 *Phys. Rev.* B **78** 144404
[32] Hikihara T, Momoi T, Furusaki A and Kawamura H 2010 *Phys. Rev.* B **81** 224433

IOP Conf. Series: Journal of Physics: Conf. Series **828** (2017) 012004 doi:10.1088/1742-6596/828/1/012004

# Impact of minute-time-scale kinetics on the stabilization of the skyrmion-lattice in $Cu_2OSeO_3$

**Johannes D Reim[1], Koya Makino[1], Daiki Higashi[1], Yusuke Nambu[2], Daisuke Okuyama[1], Taku J Sato[1], Elliot P Gilbert[3], Norman Booth[3], and Shinichiro Seki[4,5]**

[1]Institute of Multidisciplinary Research for Advanced Materials, Tohoku University, 2-1-1 Katahira, Sendai 980-8577, Japan
[2]Institute for Materials Research, 2-1-1 Katahira, Tohoku University, Sendai 980-8577, Japan
[3]Australian Centre for Neutron Scattering, Australian Nuclear Science and Technology Organization, Kirrawee DC, New South Wales 2232, Australia
[4]RIKEN Center for Emergent Matter Science (CEMS), Wako, Saitama 351-0198, Japan
[5]PRESTO, Japan Science and Technology Agency (JST), Tokyo 102-8666, Japan

E-mail: j.reim@tagen.tohoku.ac.jp

**Abstract.** Using small angle neutron scattering measurements, we have previously demonstrated the influence of field-cooling (FC), field-warming (FW) and zero-field-cooling (ZFC) protocols on the thermodynamical stability of the skyrmion-lattice phases SkX(1) and SkX(2) in $Cu_2OSeO_3$. Here, we have discovered that small variations in these protocols can have a significant impact on the phase stabilization. Using the ZFC protocol the stabilization of either SkX(1) or SkX(2) can be tuned either by applying the magnetic field directly or with intermediate steps. Furthermore, introducing waiting times at intermediate temperature steps into the FC and FW protocol significantly influences the skyrmion-lattice stabilization. Although for the latter FC and FW experiment, the extrinsic temperature equilibration effect cannot be excluded, the influence of the temperature and magnetic field sweep rate indicates the importance of slow kinetics on the order of a few minutes for the stabilization process.

## 1. Introduction

In recent years the research of chiral magnets has attracted considerable interest, as certain compounds exhibit topologically protected swirlings called skyrmions characterized by a topological quantum number [1, 2]. Besides their intriguing nature these entities offer properties well suited for application in information technology. Lattices formed from such skyrmions have been first observed in the chiral helimagnet MnSi [3] and subsequently in numerous ferromagnetic compounds such as FeGe [4], (Fe,Co)Si [5], $Cu_2OSeO_3$ [6, 7], $Co_8Zn_8Mn_4$ [8] and $GaV_4S_8$ [9]. The presence of a skyrmion lattice in $Cu_2OSeO_3$ was evidenced using small angle neutron scattering (SANS) [6, 7] and Lorentz microscopy [10]. As an insulator this compound is quite unusual among those hosting a skyrmion lattice and attracts particular interest due to the resulting multiferroic nature [10, 11]. Recently, we have reported on the thermodynamic stability of skyrmion phases in $Cu_2OSeO_3$. We were able to relate deviations concerning the stabilization of skyrmion-lattice phases, i.e., the skyrmion-lattice 1 [SkX(1)] and skyrmion-lattice 2 [SkX(1)], in phase diagrams of earlier reports to the difference in temperature ($T$) and field ($H$) protocols [12].

These skyrmion-lattice phases are related by a 30° rotation with each other [6]. Cooling the sample in an applied field (FC) stabilizes the SkX(2) phase, while SkX(1) is mostly prominent for field-warming (FW). Following a zero-field-cooling (ZFC) procedure, the SkX(1) and SkX(2) phases are stabilized in coexistence and show no sign of relaxation behavior favoring one of these phases at any point of the phase diagram. This observation indicates an intriguing phase stability most likely due to their topological nature.
Here, we investigate the different protocols in further detail. In the case of the FC and FW protocol, the temperature of the sample changes while traversing the skyrmion phase, whereas in the ZFC protocol the magnetic field is changed. In the paramagnetic (or helical/cycloidal) phase, the system is most likely independent of the path traversed by performing these changes, at least if the path is contained in the respective phase. However, in the skyrmion phase, a path-dependence is present as shown by the difference in stabilization depending on the protocol. In the following, we study the influence of introducing waiting times at intermediate steps which allow for a short time relaxation (up to four minutes) while maintaining the path itself.

## 2. Experimental

To investigate the structure of the skyrmion lattice in $Cu_2OSeO_3$, small angle neutron scattering experiments were performed on a high quality single crystal. This crystal with the approximate dimensions 8 mm × 5 mm × 3 mm has been synthesized using the chemical vapor transport method [13]. Neutron experiments were conducted at the beam line QUOKKA [14] located at the OPAL reactor of Australian Nuclear Science and Technology Organization at a wavelength $\lambda \sim 5$ Å and wavelength distribution $d\lambda/\lambda \sim 10\%$. The skyrmion lattice was stabilized using a closed-cycle $^4$He refrigerator and an external 5 T horizontal field superconducting magnet with its field aligned parallel to the incident neutron beam $\vec{k}_i$. The sample was placed on a sapphire sample plate with its [110] axis parallel to the incident neutron beam $\vec{k}_i$, and with its [001] axis along the horizontal direction. The scattering background was estimated using the SANS pattern measured at the paramagnetic temperature 60 K, and subsequently subtracted from the data shown here. The rod-shaped Bragg peaks of the skyrmion lattice enable measurement of the scattering intensity of all peaks simultaneously despite the curvature of the Ewald sphere.

## 3. Results and discussion

The two-dimensional scattering patterns measured in the SANS experiment show a set of two (helical), six [either SkX(1) or SkX(2)] or 12 Bragg reflections [coexistence and thus superposition of SkX(1) and SkX(2)] distributed at $|\vec{Q}| = Q = 0.01$ Å$^{-1}$. Figure 1(a) and (b) shows exemplary scattering patterns of superimposed skyrmion-lattice phases SkX(1) and SkX(2) stabilized using the ZFC protocol. The observed peak sets depend on the used protocol and the position in the $T$-$H$ phase diagram. In order to investigate the stabilization of SkX(1) and SkX(2) quantitatively, the scattering pattern is integrated along $\vec{Q}$ for $0.076 \leq Q \leq 0.0124$ Å$^{-1}$. The resulting pattern displays the dependence of the intensity on the azimuthal angle $\phi$. A fit function, consisting of a Gaussian function for each Bragg reflection, was used to extract the intensity, width and position of the peaks belonging to the SkX(1) and SkX(2) phases. Simultaneously, a constant term independent of $\phi$ was optimized which is related to a ring-like diffuse scattering. A detailed description of the fitting procedure can be found in [12]. While the parameters were optimized for each Friedel pair of scattering peaks individually, in the following, the intensity of SkX(1) and SkX(2) refers to the integrated intensity of all six respective peaks.

### *3.1. Impact of magnetic field sweep rate on skyrmion-lattice stabilization*

In the case of the ZFC protocol, the change of the magnetic field takes place within the skyrmion phase in contrast to the FC and FW protocols. In order to reveal a sweep rate dependence,

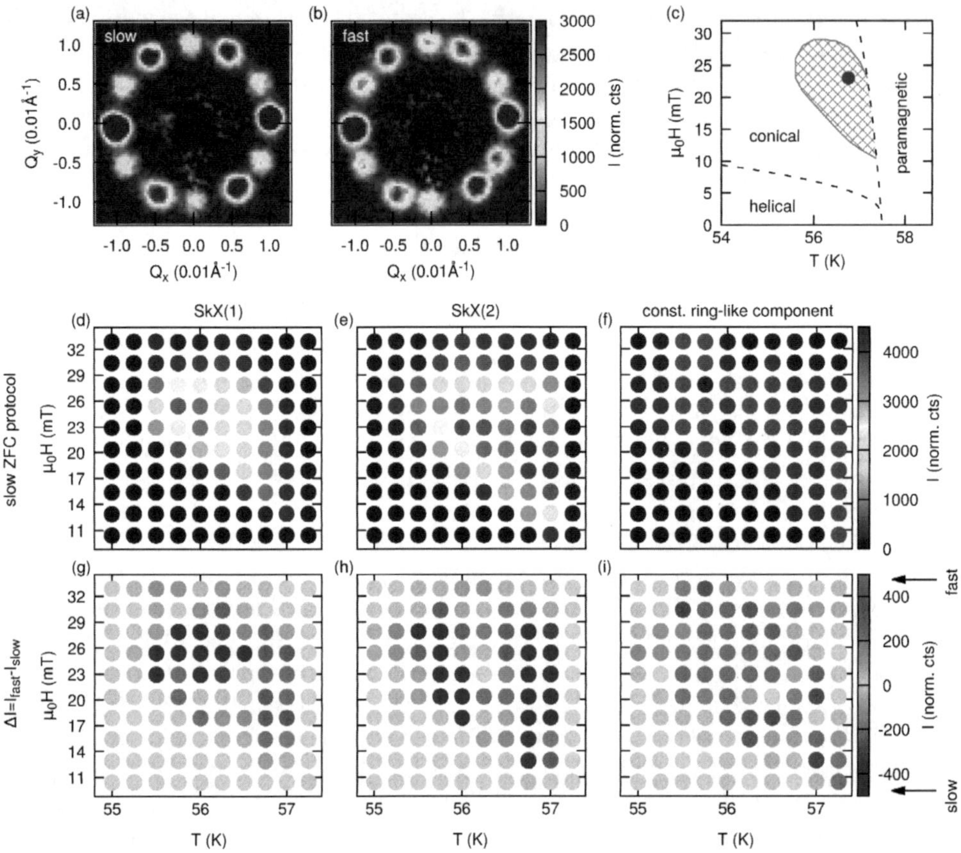


**Figure 1.** Comparison of the slow and fast ZFC protocol. The scattering pattern measured at $\mu_0 H = 23\,\mathrm{mT}$ and $T = 56.75\,\mathrm{K}$ for the slow (a) and fast (b) ZFC protocol. (c) Schematic phase diagram close to the skyrmion phase (red) and the position of the scattering patterns (a) and (b) within (blue dot). (d-f) Two-dimensional intensity maps for the SkX(1), SkX(2) and the constant ring-like component, obtained by fitting a model function to the scattering pattern. The position of each dot denotes the temperature and field of the associated measurement. (g-i) The color represents the difference in the respective intensity between the fast and slow protocol. The intensity (f) and difference (i) of the ring-like component were scaled by a factor 10.

two different ZFC protocols were used. The first one, described in [12], heats the sample up to $T = 60\,\mathrm{K}$ and cools it down to the target temperature in zero magnetic field. Subsequently, the target magnetic field is always set from the helical phase at $\mu_0 H = 0\,\mathrm{mT}$ using a rate of $23.8\,\mathrm{mT/s}$. This will here be referred to as the "fast" protocol. In the second protocol the magnetic field is increased in steps of $\mu_0 \Delta H = 2.5\,\mathrm{mT}$ starting at $\mu_0 H = 0$ up to $40\,\mathrm{mT}$. The latter protocol is described in Appendix A and will be referred to as the "slow" protocol. After reaching the target field using either of the ZFC protocols the scattering is measured for 60 s. The scattering patterns in figure 1 display skyrmion-lattices stabilized using the slow (a) and fast (b) protocol. While both show Bragg reflections of the skyrmion-lattice phases SkX(1) and dominant SkX(2), there are differences in the intensity ratio of these phases. At this point in the $T$-$H$ phase diagram [cf. figure 1(c)] the fast protocol enhances SkX(1) in favor of SkX(2). The intensity for the SkX(1), SkX(2) and ring-like diffuse component was determined by fitting the diffraction patterns measured at various $T$ and $H$. The results of the slow protocol are plotted

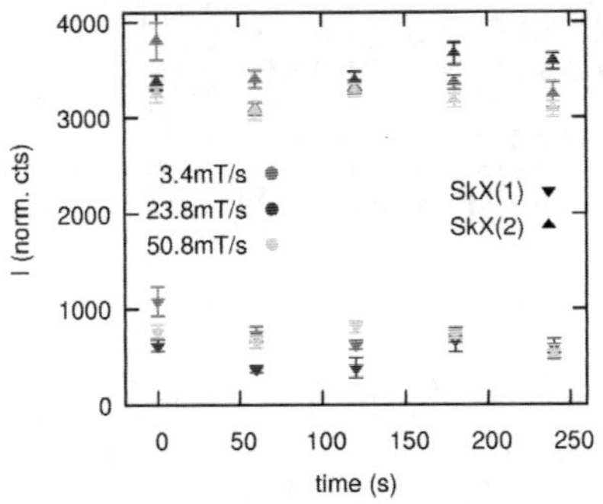


**Figure 2.** Time evolution of the integrated intensity for the SkX(1) and SkX(2) components applying the fast ZFC protocol. The protocol was performed using different rates for the increase of the magnetic field at $T = 57.0\,\mathrm{K}$. The measurement of the time evolution was started after reaching the target field $\mu_0 H = 20.4\,\mathrm{mT}$.

in 2D maps as a function of $T$ and $H$ [cf. figure 1(d-f)]. The overall structure is rather similar to the intensity maps observed for the fast protocol (see [12]), with SkX(1) being stabilized at lower-$T$ and SkX(2) being dominant in a large area of the skyrmion phase especially at higher-$T$.

Figures 1(g-i) display the difference in the integrated intensities of SkX(1), SkX(2) and the ring-like component between the fast and slow protocol. The constant component is related to short-range ordered skyrmion-lattices and a large region of the skyrmion-lattice phase (below $T = 56.75\,\mathrm{K}$ and above $\mu_0 H \sim 21\,\mathrm{mT}$) exhibits a consistently greater constant component for the fast protocol. The effect leading to the system becoming kinetically trapped is reduced for either a sufficiently slow increase of the magnetic field, or close to the phase boundary where even shorter waiting times are sufficient for the whole system to become ordered. Looking at the difference in integrated intensities of SkX(1) and SkX(2), the fast protocol stabilizes SkX(1) in favor of SkX(2) close to the skyrmion phase boundary at $T = 57.25\,\mathrm{K}$ in comparison to the slow protocol. This is quite intriguing; while SkX(2) is energetically favored close to this phase boundary, its stabilization can apparently be influenced by changing the field increase procedure. At intermediate temperatures the fast protocol stabilizes the SkX(2) slightly more compared to the slow one. However, using the slow protocol SkX(1) is stabilized more strongly in the region below $T = 56.75\,\mathrm{K}$ and above $\mu_0 H = 23\,\mathrm{mT}$, and SkX(2) below $T = 56.25\,\mathrm{K}$. Since no relaxation of the integrated intensities was observed for both skyrmion-lattices after reaching the target magnetic field, both protocols result in stable configurations. At the same time the structure of the skyrmion-lattice is the same for both protocols, as no significant deviation of the peak widths nor position in $Q$ (i.e. the extent of long-range order and its periodicity) were identified here. Furthermore, the peak sets of SkX(1) and SkX(2) both appear to retain the six-fold symmetry, neither the slow nor the fast protocol introduces a distortion to the skyrmion-lattices.

There have been recent reports on skyrmions forming a glassy structure in the vicinity of the lower-$H$ phase boundary [15, 16]; consequently the related dislocations would remain even in the long-range-ordered skyrmion-lattice phase. The increase of the constant component observed at intermediate $H$ and $T$ using the fast protocol supports the assumption of a glassy state. Furthermore, low diffuse scattering is exhibited using the slow protocol and close to the phase boundary; this corresponds with the glassy state growing into a well-ordered skyrmion-lattice. However, so far this model does not offer a compelling explanation for the $T$ and $H$-dependence of the stabilization of the SkX(1) and SkX(2) phases respectively.

As the magnetic field sweep rate plays a significant role, at $T = 57.0\,\mathrm{K}$, the fast protocol was used with three different rates (3.4, 23.8 and 50.8 mT/s) by which the magnetic field is increased to reach the target field $\mu_0 H = 20.4\,\mathrm{mT}$. With the slow protocol the target field is reached after ~640 s, which is equivalent to an average rate of 0.03 mT/s. While the difference for SkX(1) and SkX(2), in figures 1(g) and (h), is more obvious than for the chosen magnetic field rates in figure 2, the intensity difference of SkX(1) comparing the rate 3.4 mT/s to the others is of the same magnitude. This means that if even slower rates were used, the difference might

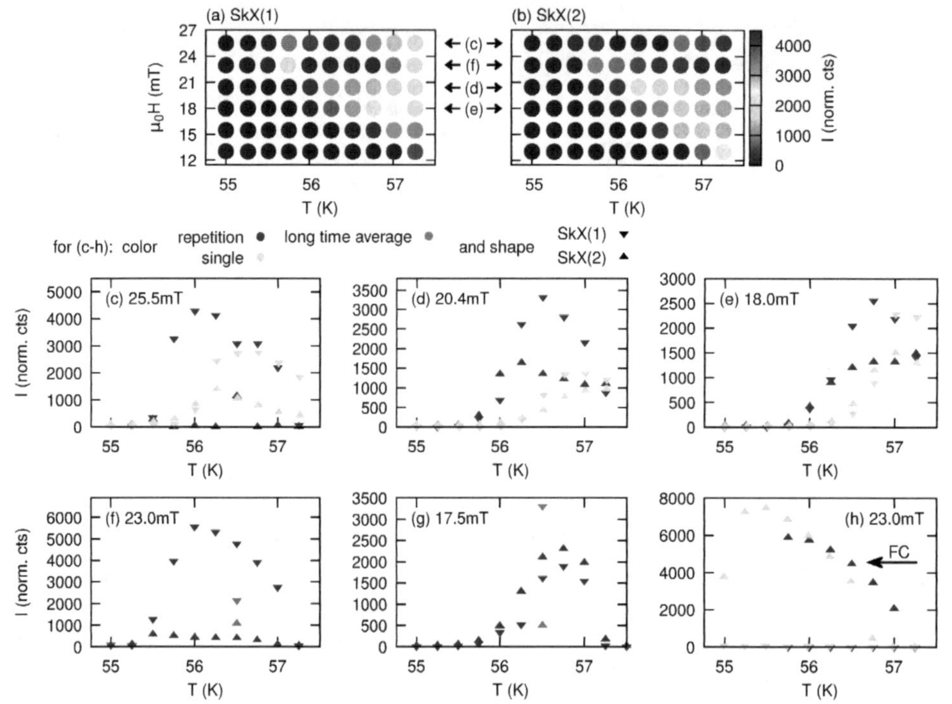


**Figure 3.** Two-dimensional intensity maps for the SkX(1) (a) and SkX(2) (b) components, obtained by fitting the model function to the azimuthal scattering pattern. Both panels represent results using FW protocols, however, at each configuration three (at $\mu_0 H = 23\,\mathrm{mT}$ five) consecutive measurements were performed, of which the first one is shown. (c-g) The integrated intensities of various constant field scans using the FW protocol. The results from single measurement scans (cyan) and the last measurement of the repetition scans (blue) are displayed for comparison. Furthermore, in (f) and (g) data points were added representing the long time average calculated (red) from time relaxation measurements. (h) For further comparison equivalent graphs obtained using the FC protocol with and without repetitions.

be enhanced further. There is a recent report on the temporal change of the skyrmion-lattice subsequent to a change of the applied magnetic field in thin-film samples of FeGe on the order of 5 s [17]. Due to the measurement time of 60 s employed here, this is not expected to play a major role.

### *3.2. Impact of temperature sweep rate: Equilibration versus intrinsic stabilization*

Ideally a similar experiment on the application of the magnetic field would have been performed for the FC and FW protocol and the influence of the temperature, too. However, our current experimental setup requires at least 600 s for the temperature equilibration for a temperature change of $\Delta T = 3\,\mathrm{K}$ [12], as the temperature sensor is mounted on the same sapphire plate at a finite distance from the sample. In order to minimize the influence from the temperature equilibration the temperature is changed in small steps of $\Delta T = 0.25\,\mathrm{K}$, starting at a temperature outside the skyrmion phase. For each set of steps, we performed one of three different repetition measurements at each temperature step: 1. a single measurement of 60 s, 2. repeat the first measurement two additional times to account for the temperature equilibration or 3. repeat the measurement four times (only used at $\mu_0 H = 17.5$ and $23\,\mathrm{mT}$). The intensity maps for

the first measurement are shown in figures 3(a-b). Comparing these maps with those of the third measurement results [12], the intensity appears slightly reduced at lower-$T$ and increased approaching the phase boundary at higher-$T$. Apart from the map's constant magnetic field scans in figures 3(c-f) an additional data set measured using the FW protocol with four repetitions at $\mu_0 H = 17.5\,\mathrm{mT}$ is shown in figure 3(g). Despite the small steps in temperature there is still a variation in intensity for each repetition measurement present. The relaxation is largest at the lower and higher-$T$ phase boundaries; however, on the basis of the available measurements, a possible intrinsic temperature dependence of the relaxation behavior cannot be distinguished from the extrinsic influence of the temperature equilibration. Comparing the measurement results in figures 3(c-e) and considering the number of repetitions indicates that longer waiting times for fields $\mu_0 H \gtrsim 18\,\mathrm{mT}$ stabilize SkX(1) over SkX(2) for the FW protocol. However, approaching the lower-$H$ phase boundary the ratio between SkX(1) and SkX(2) decreases and for $\mu_0 H = 17.5\,\mathrm{mT}$ the SkX(2) phase exhibits slightly higher intensities at all temperatures [cf. figure 3(g)]. It is important to note that longer waiting times subsequent to the stabilization of the skyrmion phases cannot make up for incomplete relaxation during the stabilization. This can be seen from the results of time relaxation measurements where the target temperature is set directly from 50 K (FW protocol). The long-time average calculated from the measurements after the temperature equilibration shows significantly different intensities for the skyrmion phases from the respective repetition measurements [cf. figures 3(f) and (g)]. Again, energetic barriers appear to prevent a relaxation of the stabilized skyrmion-lattice.
Figure 3(h) shows two data sets measured using the FC protocol, one without and the other one with four repetitions at $\mu_0 H = 23.0\,\mathrm{mT}$. The difference between these data sets is less compared to those where the FW protocol is used. When entering the skyrmion phase, the waiting times of the repetitions assist with the stabilization process of SkX(2). At lower temperatures (below $T = 56\,\mathrm{K}$), the intensity observed without repetitions exceeds the highest intensities of the other set [cf. figure 3(c)], which is peculiar when only the influence from temperature equilibration is assumed. This suggests a possible intrinsic temperature influence for the FC protocol, too.

## 4. Conclusion

Previously, we have reported on the influence of certain magnetic field and temperature protocols on the stabilization process. Here, the previously used protocols were varied slightly to investigate the influence of intermediate waiting times. The higher intensity of the ring-like component for the fast ZFC protocol indicates an increase in short-range ordered skyrmion-lattice partitions or rather in the proportion of the system in a glassy state. Yet, close to the phase boundary at $T = 57.25\,\mathrm{K}$ where the SkX(2) phase is dominant, the fast protocol strengthens the stabilization of the SkX(1) phase. The extent to which the increase in SkX(1) may be related to the freezing of the glassy state has to be studied in a future experiment. It becomes, nevertheless, clear in this work that the slow kinetics on the time-scale of a few minutes exists for the ZFC protocol. For the FC and FW protocols, the observation of the temperature equilibration complicates the study of the intrinsic temperature dependence. Based on the repetition measurements, the FC protocol appears to generally require less time for the relaxation; however, the intermediate waiting times still influence the stabilization. Using the FW protocol the system seems to require longer relaxation time. Thus, if the temperature is increased too quickly, the stabilized skyrmion phases will be different. This is even more evident in the comparison of the long-time average from the time relaxation (single step of $\Delta T = 6.5\,\mathrm{K}$) with the repetition measurements providing waiting times at each intermediate temperature step, where the influence of temperature equilibration is negligible. Here, clear differences in the stabilized skyrmion-lattice phases [cf. figures 3(f) and (g)] indicate a glass formation even more prominent compared to the fast ZFC protocol. Minute-time-scale kinetics are therefore expected to exist for the temperature sweep, too. Further studies on the influence of temperature with specifically designed experiments and optimized sample environments is highly desired. Such an approach is likely to reveal even further insight on the difference between the SkX(1) and SkX(2) phases and enable enhanced control over their respective stabilization.

**Acknowledgments**
The authors thank Y. Tokura for providing his outstanding insight into the matter at hand and N. Nagaosa for stimulating discussions. This work was partly supported by Grants-In-Aid for Scientific Research (24224009, 15H05458, 15H05883, 16H04007, and 16K13842) from MEXT of Japan. Travel expense for the experiment at QUOKKA at ANSTO was partly sponsored by the General User Program of ISSP-NSL, University of Tokyo.

## Appendix A. Details of the slow ZFC phase diagram

(i) The sample is heated up to the paramagnetic temperature ($T = 60\,\mathrm{K}$); (ii) The sample is cooled down to the target measurement temperature, the first one being $T = 57.25\,\mathrm{K}$, with the rate of 2 K/min under zero magnetic field; (iii) Wait for relatively short time (60 s) for ZFC due to beamtime restriction; (It is expected that this shorter period does not significantly change the obtained phase diagram since, while measuring the lower-field part, $0 \le \mu_0 H \le 12.5\,\mathrm{mT}$, the sample reaches its equilibrium temperature target value.) (iv) Magnetic field is set to the target magnetic field, and the SANS pattern is recorded for 1 min. The magnetic field is then set to the next target field increased with the increment of $\mu_0 \Delta H = 2.5\,\mathrm{mT}$. This step is repeated until the maximum field is reached; (v) The magnetic field is turned off ($\mu_0 H = 0$); (vi) The above steps [from (i)] are repeated with the target temperature decreased by a decrement of 0.25 K, until the lowest measurement temperature is reached.

## References

[1] Bogdanov A N and Yablonskii D A 1989 *Sov. Phys. JETP* 101
[2] Nagaosa N and Tokura Y 2013 *Nat. Nanotechnol.* **8** 899–911
[3] Mühlbauer S, Binz B, Jonietz F, Pfleiderer C, Rosch A, Neubauer A, Georgii R and Böni P 2009 *Science* **323** 915–919
[4] Yu X, Kanazawa N, Onose Y, Kimoto K, Zhang W, Ishiwata S, Matsui Y and Tokura Y 2011 *Nat. Mater.* **10** 106–109
[5] Münzer W, Neubauer A, Adams T, Mühlbauer S, Franz C, Jonietz F, Georgii R, Böni P, Pedersen B, Schmidt M, Rosch A and Pfleiderer C 2010 *Phys. Rev. B* **81**(4) 041203
[6] Seki S, Kim J H, Inosov D S, Georgii R, Keimer B, Ishiwata S and Tokura Y 2012 *Phys. Rev. B* **85** 220406
[7] Adams T, Chacon A, Wagner M, Bauer A, Brandl G, Pedersen B, Berger H, Lemmens P and Pfleiderer C 2012 *Phys. Rev. Lett.* **108** 237204
[8] Tokunaga Y, Yu X Z, White J S, Rønnow H M, Morikawa D, Taguchi Y and Tokura Y 2015 *Nat. Commun.* **6** 7638–7644
[9] Kezsmarki I, Bordacs S, Milde P, Neuber E, Eng L M, White J S, Rønnow H M, Dewhurst C D, Mochizuki M, Yanai K, Nakamura H, Ehlers D, Tsurkan V and Loidl A 2015 *Nat. Mater.* **14** 1116–1122
[10] Seki S, Yu X Z, Ishiwata S and Tokura Y 2012 *Science* **336** 198–201
[11] Seki S, Ishiwata S and Tokura Y 2012 *Phys. Rev. B* **86** 060403
[12] Makino K, Reim J D, Higashi D, Okuyama D, Sato T J, Nambu Y, Gilbert E P, Booth N, Seki S and Tokura Y 2016 *Preprint 1608.06359*
[13] Miller K H, Xu X S, Berger H, Knowles E S, Arenas D J, Meisel M W and Tanner D B 2010 *Phys. Rev. B* **82** 144107
[14] Gilbert E P, Schulz J C and Noakes T J 2006 *Physica B* **385-86** 1180–1182
[15] Rajeswari J, Huang P, Mancini G F, Murooka Y, Latychevskaia T, McGrouther D, Cantoni M, Baldini E, White J S, Magrez A, Giamarchi T, Rønnow H M and Carbone F 2015 *Proc. Nat. Acad. Sci.* **112** 14212–14217
[16] Qian F, Wilhelm H, Aqeel A, Palstra T T M, Lefering A J E, Brück E H and Pappas C 2016 *Preprint 1607.08177*
[17] Yamasaki Y, Morikawa D, Honda T, Nakao H, Murakami Y, Kanazawa N, Kawasaki M, Arima T and Tokura Y 2015 *Phys. Rev. B* **92** 220421

# Crystal Growth of a frustrated $J_1$–$J_2$ chain magnet $NaCuMoO_4(OH)$

**K Nawa**[1,2], **Y Okamoto**[1,3], **and Z Hiroi**[1]

[1] Institute for Solid State Physics, The University of Tokyo, Kashiwa, Chiba 277-8581, Japan

E-mail: knawa@tagen.tohoku.ac.jp

**Abstract.** In a frustrated $J_1$–$J_2$ chain magnet with ferromagnetic nearest neighbor interactions $J_1$ and antiferromagnetic next-nearest neighbor interactions $J_2$, novel quantum states such as a spin nematic state are theoretically expected to be present. To detect the novel states experimentally, it is required to find a suitable model compound and to obtain its sizeable and high-quality single crystals. Here we report synthesis of a single-crystalline $NaCuMoO_4(OH)$, a model compound of frustrated $J_1$–$J_2$ chain magnet, by a hydrothermal method and discuss its magnetic properties. The temperature dependencies of magnetization are almost consistent with those of powder samples at high temperatures, while the peak in heat capacity divided by temperature associated with a magnetic transition becomes sharper and larger in the single crystal. This indicates the improved quality of the single-crystalline sample compared with the powder sample. Single-crystalline $NaCuMoO_4(OH)$ should give a new experimental approach to the novel spin-nematic states expected near 26 T.

## 1. Introduction

One of interesting topics in solid state physics is to study an exotic state with an order parameter defined by a product of multiple spins [1–5]. A typical example is a spin nematic state with a order parameter $\langle s_i^+ s_{i+1}^+ s_j^- s_{j+1}^- \rangle$, where time-reversal symmetry is preserved but rotation symmetry is broken. Recently, a lot of theoretical studies have revealed that its long-range order can be lifted by competing interactions [1–5]. In this sense, a frustrated $J_1$–$J_2$ chain magnet, which contains competing ferromagnetic nearest-neighbor interactions $J_1$ and antiferromagnetic next-nearest-neighbor interactions $J_2$ in a magnetic field $B$

$$\mathcal{H} = J_1 \sum_l \mathbf{s}_l \cdot \mathbf{s}_{l+1} + J_2 \sum_l \mathbf{s}_l \cdot \mathbf{s}_{l+2} - g\mu_B B \sum_l s_l^z . \quad (1)$$

is interesting, because the spin nematic state is theoretically expected near the magnetic saturation [4 13] . However, its evidence has not been found experimentally because of some difficulties. First, detecting its subtle signatures is difficult; microscopic measurements such as NMR and neutron scattering experiments using sizeable single crystals are necessary [14–17]. Secondly, samples should have high-quality since the spin-nematic phase can be disturbed

[2] Present address: Institute of Multidisciplinary Research for Advanced Materials, Tohoku University, Sendai 980-8577, Japan

[3] Present address: Department of Applied Physics, Graduate School of Engineering, Nagoya University, Nagoya 464-8603, Japan

or its weak signatures can be masked by disorder effects. For instance, the most studied compound $LiCuVO_4$ as a candidate system to realize the spin nematic state [18–26] includes small Li-deficiency, which induces a discrepancy between NMR and magnetization data below the magnetic saturation [26]. Thus, a new $J_1$-$J_2$ chain compound that overcomes these difficulties is required.

$NaCuMoO_4(OH)$ is a new model compound of the frustrated $J_1$–$J_2$ chain magnet [27]. Exchange interactions are estimated to be $J_1 = -51$ K and $J_2 = 36$ K from magnetization and heat capacity measurements performed on a polycrystalline sample. The magnetic saturation field $B_s$ is 26 T, lower than $B_s = 41$ T for $LiCuVO_4$. The lower saturation field is advantageous for exploring the spin nematic phase by microscopic measurements in high fields. An apparent signature of a spin nematic state has not been detected in high-field magnetization measurements using polycrystalline sample [27]. However, further experiments using a single crystal would provide an evidence of the spin nematic phase, since smearing effects due to the magnetic anisotropy is removed. In this paper, we report a successful crystal growth of $NaCuMoO_4(OH)$ and discuss their magnetic properties. A sharp and large peak due to a magnetic transition observed in heat capacity measurements indicates the improved sample quality of the single crystal.

## 2. Experiments

Single crystals were grown by the hydrothermal method. First, 8.0–28.0 mmol of $MoO_3$ was dissolved in 40 ml of an aqueous solution which includes 8.0–28.0 mmol of NaOH. Then 8.0–28.0 mmol of $Cu(OH)_2$ were added and the mixed solution was stirred for a while (10–120 minutes) in contact with cold water. The solution was put in a Teflon beaker of 100 ml volume, placed in a stainless steel autoclave, and heated at 160–240 °C for 7 days. The amount of starting materials, stirring time, and heating temperature were varied to find good conditions for crystal growth. The synthetic conditions are listed in table. 1. After a heat treatment, rod-like crystals were picked up from black precipitation and washed by water. Obtained crystals were characterized by powder X-ray diffraction analysis using Cu $K_\alpha$ radiation and single crystal X-ray diffraction analysis using Mo $K_\alpha$ radiation (Rigaku, RINT-2000 and R-AXIS RAPID, respectively). Magnetic susceptibility was measured in a SQUID magnetometer (Quantum Design MPMS), and heat capacity was measured by the relaxation method (PPMS, Quantum Design).

## 3. Results and discussion

The amount of starting material, stirring time before heating, and heating temperature were tuned appropriately to obtain a single crystal. Among several synthetic conditions listed in table. 1, largest crystals were obtained when amount of starting materials, stirring time, and heating temperature were 18.2 mmol, 30 minutes, and 240 °C, respectively. Single crystals grew in the best condition is shown in Fig. 1(b). They are green in color but also include a dark-green or a black part. This is because they also include CuO and lindgrenite ($Cu_3(MoO_4)_2(OH)_2$ [28]) as inclusions. However, their amount is so small that we cannot detect them in a powder XRD pattern of powdered crystals.

When the total amount of starting materials was increased (decreased), the crystal became smaller and a lot of CuO (lindgrenite) crystals were produced after the reaction. Stirring time before heating was also crucial for the crystal growth. Figure 1 shows photographs of single crystals with different stirring time. Too short or long time of stirring makes the crystal small and increases the amount of inclusions. A surface state of $Cu(OH)_2$ particles might be important for the crystal growth.

To check the quality of the single crystal, we performed magnetization and heat capacity measurements. We discuss crystals synthesized in the best condition as judged (Fig. 1(b)) in

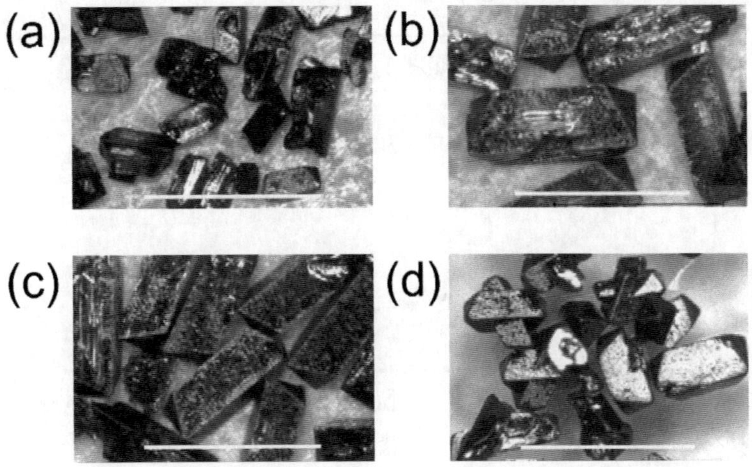


**Figure 1.** Photographs of single-crystalline $NaCuMoO_4(OH)$ grown by hydrothermal method. The solution was stirred before the heat treatment for (a) 10 minutes, (b) 30 minutes, (c) 1 hours and (d) 2 hours. Yellow scale bars indicate 1 mm.

the following. Figure 2 shows temperature dependencies of magnetization divided by magnetic field, $M/H$. Its sample-dependence is little above 20 K but large below 20 K due to an additional contribution from lindgrenite, which exhibits ferrimagnetic order at $T_N = 14$ K [28]. A magnetic field of $B = 6.614$ T is applied to reduce its contribution. Red, green, and blue solid curves represent $M/H$ for $H \parallel a$, $b$, and $c$, respectively. Their temperature dependencies are quite similar to that of the powder sample represented by a black curve, indicating that magnetic anisotropy of $NaCuMoO_4(OH)$ is small.

**Table 1.** Synthetic conditions of single-crystalline $NaCuMoO_4(OH)$. The total amount of starting materials (NaOH, $MoO_3$, and $Cu(OH)_2$) with 40 ml of water, stirring time, and heating temperature, and a typical crystal size are listed. The optimized condition is emphasized by bold text.

| amount (mmol) | stirring time (min) | temperature (°C) | typical size (mm) |
|---|---|---|---|
| 8.0 | 30 | 240 | – |
| 12.0 | 30 | 240 | – |
| 16.0 | 30 | 240 | 0.3×0.7×0.3 |
| 28.0 | 30 | 240 | 0.2×0.5×0.2 |
| 18.2 | 10 | 240 | 0.1×0.3×0.1 (Fig. 1(a)) |
| **18.2** | **30** | **240** | **0.4×1.0×0.4 (Fig. 1(b))** |
| 18.2 | 60 | 240 | 0.3×0.7×0.3 (Fig. 1(c)) |
| 18.2 | 120 | 240 | 0.2×0.5×0.2 (Fig. 1(d)) |
| 8.0 | 30 | 160 | – |
| 18.2 | 30 | 160 | – |

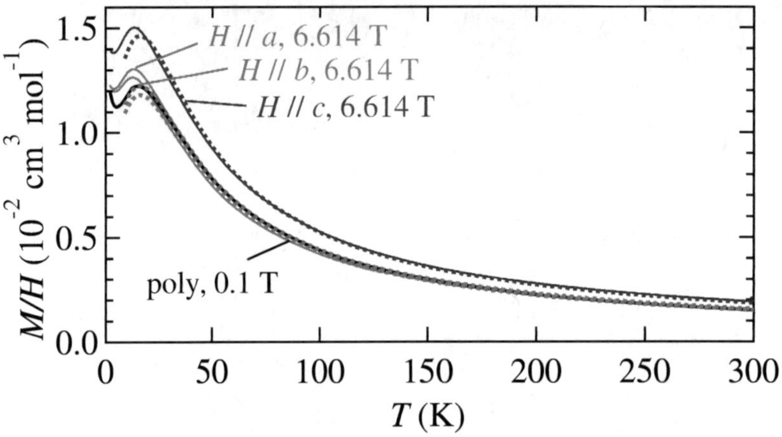


**Figure 2.** Temperature dependencies of magnetization divided by magnetic field, for $B \parallel a$, $b$, and $c$ in $B = 6.614$ T. Dashed curves stand for a simulated curve of the frustrated $J_1$–$J_2$ chain with $J_1 = -51$ K and $J_2 = 36$ K. Black solid curve indicates a temperature dependence of magnetic susceptibility of powder samples at 0.1 T.

The $M/H$ at 20–300 K is reproduced by the following function,

$$M/H = \frac{N g_i^2 \mu_B^2}{k_B J_2} \chi_{\text{chain}}(T) + \chi_0 \quad (i = a, b, c), \tag{2}$$

where $N$, $g_i$, $\chi_{\text{chain}}$, and $\chi_0$ indicate the Avogadro constant, diagonal components of $g$-tensor, magnetic susceptibility of frustrated $J_1$–$J_2$ chain, and temperature independent magnetic susceptibility, respectively. $\chi_{\text{chain}}$ is calculated for $J_1 = -1.4$ and $J_2 = 1$ by an exact diagonalization technique using ALPS package [29]. As shown in Fig. 2, experimental and simulated curves show good agreement. The best fit is obtained for $g_a = 2.15$, $g_b = 2.11$, $g_c = 2.35$, and $\chi_0 = 9.71 \times 10^{-5}$ cm$^3$ mol$^{-1}$. From the deviation of simulated curves below 15 K, the amount of lindgrenite is estimated to be less than 1 % in a molar quantity.

The improved quality of the single crystal is confirmed by heat capacity measurements. Figure 3 shows temperature dependencies of heat capacity divided by temperature, $C/T$, of a few single-crystalline and powder samples. A lattice contribution in $C/T$ is negligible and smaller than 0.2% of the total $C/T$ in this temperature range. Powder sample 2 is the same sample used in ref. [27], and powder sample 1 is a new sample prepared hydrothermally from 40 mmol of $Na_2MoO_4 \cdot 2H_2O$, 1.912 mmol of $Cu(NO_3)_2 \cdot 3H_2O$, and 10 ml of water. For all the samples $C/T$ exhibits a peak at about 0.6 K due to a magnetic transition. The peak apparently becomes sharp and large for the single-crystalline samples compared with the powder samples. Such a large difference in the peak magnitude indicates that small chemical disorder makes the magnetic transition diffusive, as is often the case with quasi-1D compounds. Single-crystalline sample 1 should have the best quality among the samples. The transition temperature $T_N$ is determined to be 0.63(1) K from the peak position, which is slightly higher than 0.59 K for the powder sample 2 [27]. The rough estimate of magnetic entropy $S_m$ below $T_N$ gives 0.17 J mol$^{-1}$, assuming that $C/T$ decreases to 0 by $T^3$ shown as a dashed curve in Fig. 3.

Now single-crystalline $NaCuMoO_4(OH)$ with improved quality becomes available. It would enable further studies using a high magnetic field, such as magnetization and NMR experiments. We are now planing high field NMR experiments using a static magnetic field to elucidate the nature of the spin nematic phase. In addition, neutron scattering experiments are also possible

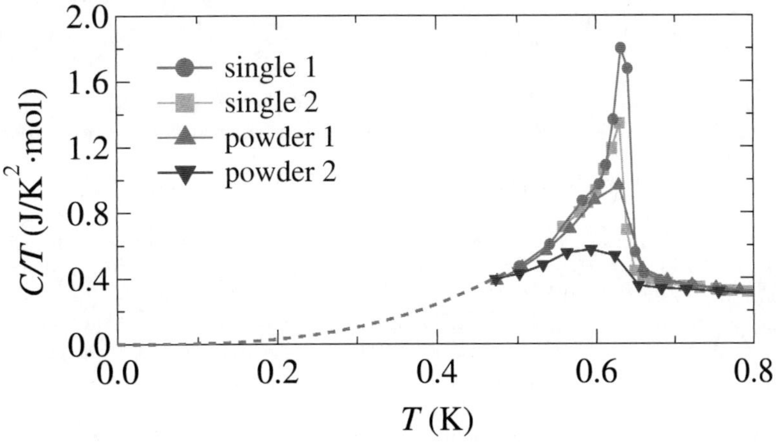


**Figure 3.** Temperature dependencies of heat capacity divided by temperature, $C/T$, in zero field. $C/T$ of two single-crystalline and two powder samples are compared. A dashed curve is assumed to estimate magnetic entropy $S_m$ of the single-crystalline sample 1.

by using a pulse field [30] or a static field [31, 32] owing to the saturation field of 26 T lower than that of $LiCuVO_4$. We believe that these experiments give a new insight for the novel spin-nematic state.

## 4. Summary

We succeeded in synthesizing single crystals of $NaCuMoO_4(OH)$, a model compound of a frustrated $J_1$–$J_2$ chain magnet. The optimized condition for the crystal growth is reported. The magnetic transition of the single-crystalline samples is much sharper than that of powder samples, indicating their improved quality, while at high temperatures magnetic properties of both samples are almost consistent. Further experiments using single-crystalline $NaCuMoO_4(OH)$ will uncover the physics of the $J_1$–$J_2$ chain magnet.

## Acknowledgments

We thank H. Ishikawa, M. Yoshida, T. Masuda, S. Asai, and T. Oyama for fruitful discussions. This work was supported by Grant-in-Aid for Young Scientists (B) (No. 15K17693).

## References

[1] Andreev A F and Grishchuk I A 1984 *Zh. Eksp. Teor. Fiz.* **87**, 467–75.
[2] Shannon N, Momoi T, and Sindzingre P 2006 *Phys. Rev. Lett.* **96**, 027213.
[3] Momoi T, Sindzingre P, and Shannon N 2006 *Phys. Rev. Lett.* **97** 257204.
[4] Ueda H T and Momoi T 2013 *Phys. Rev. B*, **87**, 144417.
[5] Zhitomirsky M E and Tsunetsugu H 2010 *Europhys. Lett.* **92**, 37001.
[6] Chubukov A V 1991 *Phys. Rev. B* **44**, 4693–96.
[7] Kecke L, Momoi T, and Furusaki A 2007 *Phys. Rev. B* **76** 060407.
[8] Vekua T, Honecker A, Mikeska H -J, and Heidrich-Meisner F 2007 *Phys. Rev. B* **76**, 174420.
[9] Hikihara T, Kecke L, Momoi T, and Furusaki A 2008 *Phys. Rev. B* **78**, 144404.
[10] Sudan J, L'uscher A, and L'auchli A M 2009 *Phys. Rev. B* **80**, 140402.
[11] Sato M, Hikihara T, and Momoi T 2013 *Phys. Rev. Lett.* **110**, 077206.
[12] Starykh O A and Balents L 2014 *Phys. Rev. B* **89** 104407.
[13] Balents L and Starykh O A 2016 *Phys. Rev. Lett.* **116** 177201.
[14] Sato M, Momoi T, and Furusaki A 2009 *Phys. Rev. B* **79**, 060406.
[15] Sato M, Hikihara T, and Momoi T 2011 *Phys. Rev. B* **83**, 064405.

[16] Smerald A and Shannon N 2013 *Preprint* cond-mat/1303.4465v1; Smerald A and Shannon N 2016 *Phys. Rev. B* **93**, 184419.
[17] Onishi H 2015 *J. Phys. Soc. Jpn.* **84**, 083702.
[18] Lafontaine M A, Leblanc M, and Ferey G 1989 *Inorg. Chem.* **C45**, 1205–6.
[19] Enderle M, Mukherjee C, Fåk B, Kremer R K, Broto J -M, Rosner H, Drechsler S -L , Richter J, Malek J, Prokofiev A, Assmus W, Pujol S, Raggazzoni J -L, Rakoto H, Rheinst́adter M and Rønnow H M 2005 *Europhys. Lett.* **70**, 237–43.
[20] Büttgen N, Krug von Nidda H -A, Svistov L E, Prozorova L A, Prokofiev A, and Aßmus W 2007 *Phys. Rev. B* **76**, 014440.
[21] Masuda T, Hagihala M, Kondoh Y, Kaneko K, and Metoki N 2011 *J. Phys. Soc. Jpn.* **80**, 113705.
[22] Mourigal M, Enderle M, Fåk B, Kremer R K, Law J M, Schneidewind A, Hiess A, and Prokofiev A 2012 *Phys. Rev. Lett.* **109**, 027203.
[23] Büttgen N, Kraetschmer W, Svistov L E, Prozorova L A, and Prokofiev A 2010 *Phys. Rev. B* **81**, 052403.
[24] Nawa K, Takigawa M, Yoshida M, and Yoshimura K, 2013 *J. Phys. Soc. Jpn.* **82**, 094709.
[25] Svistov L E, Fujita T, Yamaguchi H, Kimura S, Omura K, Prokofiev A, Smirnov A I, Honda Z, and Hagiwara M 2011 *JETP Lett.* **93**, 21–5.
[26] Büttgen N, Nawa K, Fujita T, Hagiwara M, Kuhns P, Prokofiev A, Reyes A P, Svistov L E, Yoshimura K, and Takigawa M 2014 *Phys. Rev. B* **90**, 134401.
[27] Nawa K, Okamoto Y, Matsuo A, Kindo K, Kitahara Y, Yoshida S, Ikeda S, Hara S, Sakurai T, Okubo S, Ohta H, and Hiroi Z 2014 *J. Phys. Soc. Jpn.* **83**, 103702.
[28] Vilminot S, Andre G, Richard-Plouet M, Bourée-Vigneron F, and Kurmoo M 2006 *Inorg, Chem.* **45**, 10938–46.
[29] Bauer B, Carr L D, Evertz H G, Feiguin A, Freire J, Fuchs S, Gamper L, Gukelberger J, Gull E, Gurtler S, Hehn A, Igarashi R, Isakov S V, Koop D, Ma P N, Mates P, Matsuo H, Parcollet O, Pawlowski G, Picon J D, Pollet L, Santos E, Scarola V W, Schollwöck U, Silva C, Surer B, Todo S, Trebst S, Troyer M, Wall M L, Werner P and Wessel S 2011 *J. Stat. Mech.* P05001; Albuquerque A F, Alet F, Corboz P, Dayal P, Feiguin A, Fuchs S, Gamper L, Gull E, Gürtler S, Honecker A, Igarashi R, Körner M, Kozhevnikov A, Läuchli A, Manmana S R, Matsumoto M, McCulloch I P, Michel F, Noack R M, Pawlowski G, Pollet L, Pruschke T, Schollwöck U, Todo S, Trebst S, Troyer M, Werner P, Wessel S 2007 *J. Mag. Mag. Mater.* **310**, 1187–93.
[30] Nojiri H, Yoshii S, Yasui M, Okada K, Matsuda M, Jung J -S, Kimura T, Santodonato L, Granroth G E, Ross K A, Carlo J P, and Gaulin B D 2011 *Phys. Rev. Lett.* **106** 237202.
[31] Smeibidl P, Tennant A, Ehmler H, and Bird M 2010 *J. Low Temp. Phys.* **159**, 402–5.
[32] Smeibidl P, Bird M, Ehmler H, Dixon I, Heinrich J, Hoffmann M, Kempfer S, Bole S, Toth J, Prokhnenko O, and Lake B, 2016 *IEEE Trans. Appl. Supercond.* **26**, 4301606.

# Powder neutron diffraction in one-dimensional frustrated chain compound $NaCuMoO_4(OD)$

**S Asai[1], T Oyama[1], M Soda[1], K Rule[2], K Nawa[3], Z Hiroi[1], and T Masuda[1]**

[1]Institute for Solid State Physics, The University of Tokyo, Kashiwanoha, Kashiwa, Chiba 277-8581, Japan
[2]Bragg Institute, Australian Nuclear Science and Technology Organization, Lucas Heights, New South Wales 2234, Australia
[3]Institute of Multidisciplinary Research for Advanced Materials, Tohoku University, Sendai, Miyagi 980-8577, Japan

E-mail: asai.shinichiro@issp.u-tokyo.ac.jp

**Abstract.** We carried out powder neutron diffraction measurements in $NaCuMoO_4(OD)$ in order to investigate the ground state of this material. We observe no temperature dependence between the powder neutron diffraction profiles at $T = 1.2$ and 0.07 K within the experimental error. We evaluate the maximum of the expected moment under the assumption that the helical magnetic order is realized and that the magnetic peaks are hidden in background scattering. We determined the positions of the D ions from the profile by using the Rietveld analysis.

## 1. Introduction

In $S = 1/2$ one-dimensional frustrated chain with nearest-neighbor ferromagnetic interaction $J_1$ and next-nearest-neighbor antiferromagnetic interaction $J_2$, the ground state exhibits various quantum phases depending on the ratio between $J_1$ and $J_2$ [1]. Theoretical study predicted that applying magnetic field induces exotic magnetic state, such as the spin multipole order in the vicinity of the saturation field [2]. However, such a state has not been well identified because of high saturation field and/or difficulty of obtaining a large crystal for the existing model compounds. For a famous model compound $LiCuVO_4$, the neutron scattering study suggests that the helical magnetic structure is the ground state in zero magnetic field [3]. The spin-flop transition [4] and the field-induced spin density wave state [5] were investigated by the neutron diffraction in the magnetic field. The magnetization-field curve indicated the presence of the unconventional magnetic state just below the saturation field of 44.4 T [6], but the field was too high to identify the spin correlation using neutron scattering technique.

Recently a new model compound $NaCuMoO_4(OH)$ was reported [7]. The crystal structure [8] is shown in Fig. 1. $Cu^{2+}$ ions carrying S = 1/2 form one-dimensional chain along the crystallographic $b$ direction. The magnetic transition temperature of 0.6 K, the saturation field of 26 T, and the synthesis of a large single crystal were reported [7]. The field is covered by pulse magnet for neutron scattering [10] or hybrid magnet developed in Helmholz Zentrum Berlin [11]. Thus this compound is a good candidate for studying the quantum phases in the magnetic field. The ratio between $J_1$ and $J_2$ is evaluated to be -1.4 from the analysis

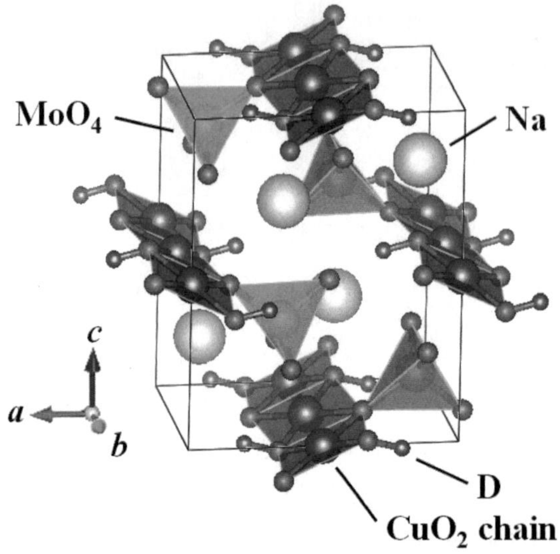


**Figure 1.** Crystal structure of $NaCuMoO_4(OD)$. The positions of the D ions are determined by this study. VESTA software [9] is used for drawing crystal structure.

of the magnetic susceptibility, which indicates that a helical magnetic order is realized at low temperature.

In this study, we performed neutron diffraction experiments in order to investigate the ground state of $NaCuMoO_4(OD)$, the D isotope for $NaCuMoO_4(OH)$. We observed no magnetic peak within the experimental error. We estimated the the moment size under the assumption that the helical magnetic order is realized and that the incommensurate magnetic peaks are hidden in background scattering. In addition we determined the position of the D ions from the diffraction profile by using the Rietveld analysis.

## 2. Experiments

The deuterated polycrystalline sample was synthesized by hydrothermal method. Neutron diffraction experiments were performed on thermal neutron triple-axis spectrometer TAIPAN at OPAL reactor, ANSTO [12]. The collimation setting was open - 40′ - 40′ - open. The PG filter was located in front of the analyzer. The incident neutron energy of 14.87 meV was used for the experiments. We used a dilution cryostat to achieve the temperature of 0.07 K. The deuterated isopropanol was put in the powder sample for improving the thermal conductivity of the sample. The obtained neutron patterns were analyzed by the Rietveld method using the FULLPROF software [13]. For simplicity, we fixed the occupancy parameters to be 1 and the isotropic temperature parameters to be 0 during the structure analysis.

## 3. Results and Discussions

Figure 2 shows the powder neutron diffraction pattens at $T$ = 0.07 and 1.2 K. The magnetic peaks had been expected at $2\theta$ = 16.7 and 24.2 deg. at 0.07 K in the case that the helical magnetic order was induced by the magnetic frustration with $J_1/J_2$ = -1.4 [7]. We, however, did not observe any difference between the profiles and did not detect any magnetic peak. It is due to the high background intensity from incoherent scattering, which reduces the S/N ratio of the diffraction profile. We, then, presume that a magnetic signal is hidden in the background, and evaluate the maximum of the moment $m$, at best, by comparing an expected magnetic signal

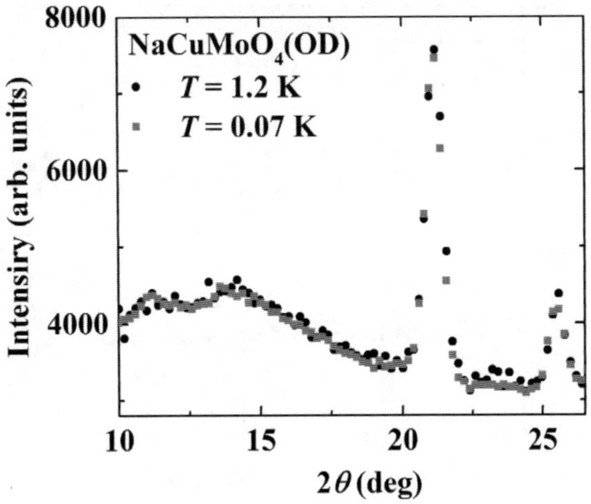


**Figure 2.** Powder neutron diffraction patterns at $T$ = 0.07 and 1.2 K. Data are vertically scaled.

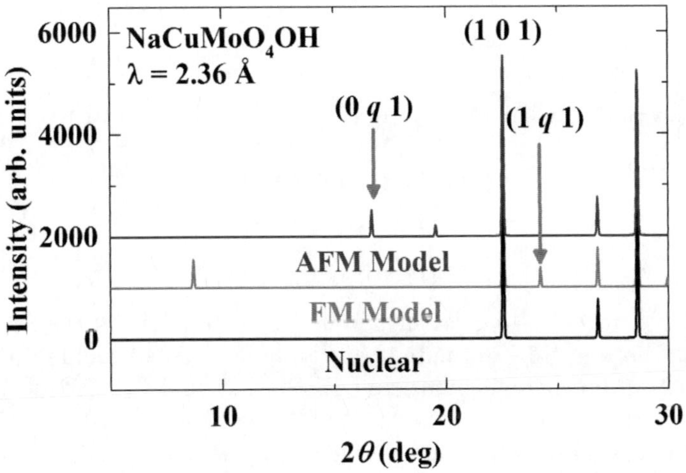


**Figure 3.** Calculated powder neutron diffraction patterns. The details of the AFM and FM models are shown in the main text.

with the statistical error of the background intensity. Here we assume that the spiral plane of the helical magnetic structure is in the crystallographic $ab$ plane. We consider two cases; one is antiferromagnetically coupled chains, AFM model, and another is ferromagnetically coupled chains, FM model.

We first calculate the neutron diffraction profile in the case of $m = 1\ \mu_B$ as shown in Fig. 3. In this paper, we does not consider the $2\theta$ range below 10 deg. because this range is not covered by the TAIPAN spectrometer. The position, index number, and ratio of the intensity on the (101) nuclear peak for the largest magnetic peak are 16.4 deg., (0 $q$ 1), and 0.149 for the AFM model, and 24.2 deg., (1 $q$ 1), and 0.105 for the FM model in our calculation, where $q$ is equal to 0.3862. Since the integrated intensity of the magnetic peak is proportional to $m^2$, the

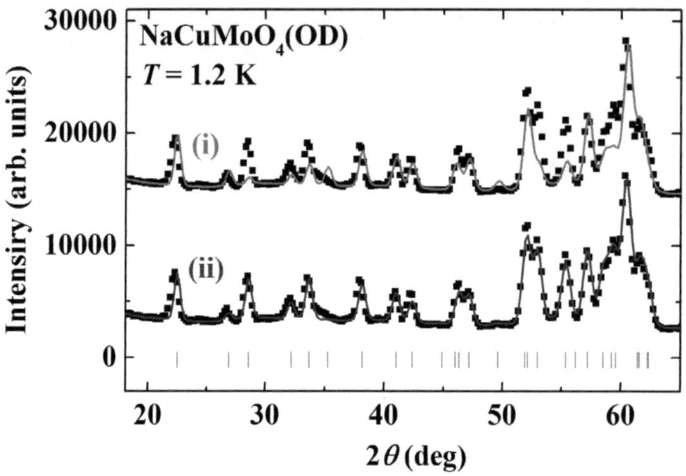


**Figure 4.** Powder neutron diffraction pattern at $T = 1.2$ K. The closed squares represent the experimentally obtained data. The upper and lower solid curves represent the calculated curves. The vertical bars below the closed squares represent the position of the nuclear Bragg peaks for $NaCuMoO_4(OD)$.

intensities of the magnetic peaks are written as

$$I(0q1) = 0.149m^2 I(101) \tag{1}$$

and

$$I(1q1) = 0.105m^2 I(101) \tag{2}$$

for the AFM and FM models, respectively. Here $I(hkl)$ represents the intensity of the peak indexed as $(hkl)$. We roughly evaluated $I(101)$ to be 10000 counts from the fitting the experimental data to the Gaussian function. The background intensities and their statistical errors at 16.7 and 24.2 deg. are about $12000 \pm 109$ and $10000 \pm 100$ counts, respectively. We substitute all of the evaluated value and derived equation shown above for the relation that the statistical error of the background intensity is larger than $I(0q1)$ and $I(1q1)$, and finally evaluated the ordered moment $m$ to be below 0.27 and 0.31 $\mu_B$ for AFM and FM models, respectively. The neutron diffraction study by using the single crystal is needed for further investigation, which is now in progress.

Figure 4 shows the powder neutron diffraction pattens at $T = 1.2$ K. We can index all of the peaks by the orthorhombic structure with the space group $Pnma$ except for the aluminum one at 61 deg. The obtained lattice parameters are $a = 7.741(3)$, $b = 9.495(4)$, and $c = 5.960(2)$ Å. The solid curve (i) represents the calculated diffraction profile by using the structure model reported in Ref. 8. It does not reproduce the experimental data. We should note that the hydrogen ions were not taken into consideration in the previous structural analysis [8]. In fact, neutron diffraction profiles are much more affected by the presence of the hydrogen/deuteron ions than x-ray diffraction profiles are. Then we perform Rietveld analysis including the D ions. In order not to conflict with the chemical composition ratio, we assume that the D ions are located in special position denoted by $4a$, $4b$, or $4c$ in the crystal structure with the space group $Pnma$. We exclude the $4b$ Wyckoff position (0, 0, 1/2) because $Cu^{2+}$ ions occupy it. Firstly, we try the case that the D ions are in the $4a$ Wyckoff position (0, 0, 0), but the calculated profile does not reproduce the data. Secondly we try the $4c$ Wyckoff position ($x$, 1/2, $z$), and fit to the data is reasonable as shown in the blue curve in Fig. 4. The position of the D ions is determined to be (0.243(4), 0.25, 0.030(4)) and is illustrated in Fig. 1.

## 4. Summary

We carried out powder neutron diffraction measurements for $NaCuMoO_4(OD)$ in order to investigate the ground state and crystal structure. We observed no temperature dependence between the profiles at $T = 1.2$ and 0.07 K within the experimental error. By comparing the expected magnetic signal with the statistical error, we evaluated the ordered moment to be less than 0.31 $\mu_B$. In addition we determined the position of the D ions to be (0.243(4), 0.25, 0.030(4)).

## Acknowledgments

Travel expenses for the neutron diffraction experiment performed using TAIPAN at ANSTO, Australia, were supported by the General User Program for Neutron Scattering Experiments, Institute for Solid State Physics, The University of Tokyo (Proposal No. 15628), at JRR-3, Japan Atomic Energy Agency, Tokai, Japan.

## References

[1] Furukawa S, Sato S M and, Onoda S 2010 *Phys. Rev. Lett.* **105** 257205.
[2] Hikihara T, Kecke L, Momoi T and Furusaki A 2008 *Phys. Rev. B* **78** 144404.
[3] Gibson B J, Kremer R K, Prokofiev A V, Assmus W, and McIntyre G J 2004 *Physica B* **350** e253.
[4] Yasui Y, Naito Y, Sato K, Moyoshi T, Sato M, and Kakurai K 2008 *J. Phys. Soc. Jpn.* **77** 023712.
[5] Masuda T, Hagihala M, Kondoh Y, Kaneko K, and Metoki N 2011 *J. Phys. Soc. Jpn.* **80** 113705.
[6] Svitov L E, Fujita T, Yamaguchi H, Kimura S, Omura K, Prokofiev A, Smirnov A I, Honda Z, and Hagiwara M 2011 *JETP Lett.* **93** 21.
[7] Nawa K, Okamoto Y, Matsuo A, Kindo K, Kitahara Y, Yoshida S, Ikeda S, Hara S, Sakurai T, Okubo S, Ohta H, and Hiroi Z 2014 *J. Phys. Coc. Jpn.* **83** 103702.
[8] Moini A, Peascoe R, Rudolf P R, and Clearfield A 1986 *Inorg. Chem.* **25** 3782.
[9] Momma K and Izumi F 2011 *J. Appl. Crystallogr.* **44** 1272.
[10] Nojiri H, Yoshii S, Yasui M, Okada K, Matsuda m, Jung J -S, Kimura T, Santodonato L, Granroth G E, Ross K A, Carlo J P, and Gaulin B D 2011 *Phys. Rev. Lett.* **106** 237202.
[11] Prokhnenko O, Stein W -D, Bleif H -J, Fromme M, Bartkowiak M, and Wilpert T 2015 *Rev. Sci. Instr.* **86** 033102.
[12] Danilkin S A, Horton G, Moore R, Braoudakis G, and Hagen M E 2007 *J. Neutron Res.* **15** 55.
[13] Rodriguez-Carvajal J 1993 *Physica B* **192** 55.

IOP Conf. Series: Journal of Physics: Conf. Series **828** (2017) 012007 doi:10.1088/1742-6596/828/1/012007

# Thermodynamic properties of quadrupolar states in the frustrated pyrochlore magnet $Tb_2Ti_2O_7$

**H. Takatsu[1,2], T. Taniguchi[2], S. Kittaka[3], T. Sakakibara[3], and H. Kadowaki[2]**

[1]Department of Energy and Hydrocarbon Chemistry, Graduate School of Engineering, Kyoto University, Kyoto 615-8510, Japan
[2]Department of Physics, Tokyo Metropolitan University, Hachioji-shi, Tokyo 192-0397, Japan
[3]Institute for Solid State Physics, University of Tokyo, Kashiwa 277-8581, Japan

**Abstract.** The low-temperature thermodynamic properties of the frustrated pyrochlore $Tb_{2+x}Ti_{2-x}O_{7+y}$ have been studied using the single crystal of $x = 0.005$ sitting in a long range ordered phase in the $x$–$T$ phase diagram. We observed that the specific heat exhibits a minimum around 2 K and slightly increases on cooling, similar to a Schottky-like anomaly for canonical spin ices. A clear specific-heat peak observed at $T_c = 0.53$ K is ascribable to the phase transition to a quadrupolar state, which contributes to a relatively large change in entropy, $S \simeq 2.7$ J K$^{-1}$mol$^{-1}$. However, it is still smaller than $R\ln 2$ for the ground state doublet of the Tb ions. The entropy release persists to higher temperatures, suggesting strong fluctuations associated with spin ice correlations above $T_c$. We discuss the field dependence of the entropy change for $H||[111]$ and $H||[001]$.

## 1. Introduction

Geometrically frustrated magnets have attracted much attention because of the realization of new type of electronic and magnetic phenomena with unconventional order parameters [1, 2]. In particular, the pyrochlore-lattice magnet $Tb_{2+x}Ti_{2-x}O_{7+y}$, a putative candidate of quantum spin liquid (QSL) [3, 4], shows unique properties including an unknown long range order (LRO) in the vicinity ($x_c = -0.0025 \leq x < 0.04$) of the QSL state [5, 6]. Indeed a clear specific-heat peak was observed at $T_c \simeq 0.5$ K for the sample with $x = 0.005$, while no LRO associated with the large magnetic and/or structural phase transitions was confirmed [5]. Although the only small Bragg peak with the order of 0.1 $\mu_B$/Tb appears below $T_c$, it is too small to explain the corresponding entropy change in the specific heat. It is thus apparently different from the magnetic dipole order inferred by earlier theories [7, 8]. This mysterious, or hidden, order is an important subject for the study of the actual nature of the ground state of $Tb_2Ti_2O_7$ [9–18].

Recently, we have investigated the hidden order of $Tb_{2+x}Ti_{2-x}O_{7+y}$ using a single crystalline sample with $x = 0.005$ ($T_c = 0.53$ K) by means of neutron scattering, specific heat, magnetization measurements [19, 20]. From the semi-quantitative analysis based on the theoretical model proposed by Onoda and Tanaka [21, 22], we have demonstrated that the ordered state originates from electric quadrupole moments inherent in the non-Kramers ion of $Tb^{3+}$. It is remarkable that the estimated parameter set is located very close to the phase boundary between the quadrupolar and U(1) QSL states. This result naturally explains the previous experimental result that the minute change in $x$ induces a phase transition between the QSL and LRO states [5, 6]. These results also showed remarkable behaviors and possibilities for magnetic field, such as a two dimensional (2D) quadrupole order for $H||[111]$, where the system behaves as decoupled 2D kagomé layers of quadrupole moments separated by triangular layers of polarized magnetic moments, which is reminiscent of the so-called kagomé ice (KI) state of spin ice (SI) materials [23–29].

Therefore, it is intriguing to examine thermodynamic properties of the quadrupolar state in $Tb_{2+x}Ti_{2-x}O_{7+y}$ under magnetic field. For this purpose, we studied specific heat ($C_P$) and the entropy ($S$) change of a sample showing the LRO. Here we focus on experiments in the [111] and [001] field directions and show the $T$– and $H$–dependence of $S$ and $\Delta S[= S(0.55\mathrm{K}, H) - S(0.15\mathrm{K}, H)]$. We found that the plateau-like behavior of $\Delta S$ appears in fields around 0.5 T only for $H||[111]$, while the change is soon suppressed for $H||[001]$. This plateau state is attributed to the change in states or the formation of the quadrupole order on the kagomé layers perpendicular to the magnetic field.

## 2. Experimental

Single crystals of $Tb_{2+x}Ti_{2-x}O_{7+y}$ were grown by a floating zone method [6]. We used the crystal with $x = 0.005$. Specific heat $C_P$ was measured by a quasi-adiabatic heat-pulse or thermal relaxation method. In a temperature range below 2 K, we used a dilution, $^3$He, and an adiabatic demagnetization refrigerators, while above 2 K we used a Quantum Design PPMS system. The in-field data presented here were obtained using a vector magnet system where an accuracy of the field direction to the sample is below 1°. In order to reduce the demagnetization effect, we used a plate-like crystal along the $\langle 110 \rangle$ plane which includes the [111], [110], and [001] axes. The sample is approximately $0.7 \times 0.9 \times 0.1$ mm$^3$ which is 0.35 mg in weight. Since the demagnetization factor for the [111] and [001] directions is small enough ($N \simeq 0.09$), demagnetization corrections were not performed in the present study.

## 3. Results and Discussion

Temperature dependence of $C_P$ at zero field is shown in Fig. 1. The specific heat $C_P$ exhibits a minimum around 2 K and slightly increases on cooling toward $T_c = 0.53$ K, implying a Schottky-type anomaly characterized by SI correlations [30]. A clear peak at $T_c$ results in the phase transition to the quadrupolar state [20]. These behaviors are compatible with the experimental results of the polycrystalline sample of $x = 0.005$ [5].

In order to analyze the low-$T$ behavior of the specific heat of the ground state doublet ($C_{\mathrm{GND}}$) and its entropy change, it is important to estimate other contributions and subtract those from the measured $C_P$ data. The specific heat of insulating $Tb_{2+x}Ti_{2-x}O_{7+y}$ is represented as $C_P = C_{\mathrm{GND}} + C_{\mathrm{CEF}} + C_{\mathrm{L}} + C_{\mathrm{N}}$. Here $C_{\mathrm{CEF}}$ is a contribution attributed to higher-energy crystal electric field (CEF) states. This contribution was calculated by taking the CEF scheme obtained by Ref. [31] and assuming the Schottky specific heat. It is slightly visible at temperatures above 2 K and becomes negligible below 1 K. $C_{\mathrm{L}}$ is the lattice specific heat estimated in the same way as described in Ref. [7]. As reference, this is negligibly small at low temperatures (viz. below 3 K) [7, 32]. $C_{\mathrm{N}}$ is the nuclear specific heat showing the Schottky anomaly. It is finite in a LRO state with the finite magnetic-dipole or electric-quadrupole hyperfine field from $4f$ moment [33, 34] or in external magnetic field [35], because of the nuclear level splitting of the nuclear spin $I = 3/2$ of $^{159}$Tb with the natural abundance 100%, which mostly contributes to $C_{\mathrm{N}}$. For this estimation, we fitted the low-$T$ part in the range between 0.1 and 0.4 K using the relation of $C_P \simeq C_{\mathrm{GND}} + C_{\mathrm{N}} = AT^n + B/T^2$, where $C_{\mathrm{GND}}$ is assumed to be the power law temperature dependence with respect to low-$T$ excitations for the phase transition, and $C_{\mathrm{N}}$ is tentatively used as the term proportional to $1/T^2$: $A$ and $B$ are the coefficients of these $T$ dependences. One of the fitting results is shown in Fig. 1. The low-$T$ behavior down to 0.15 K can be fitted by this simple relation. However, it is not applicable for the data below

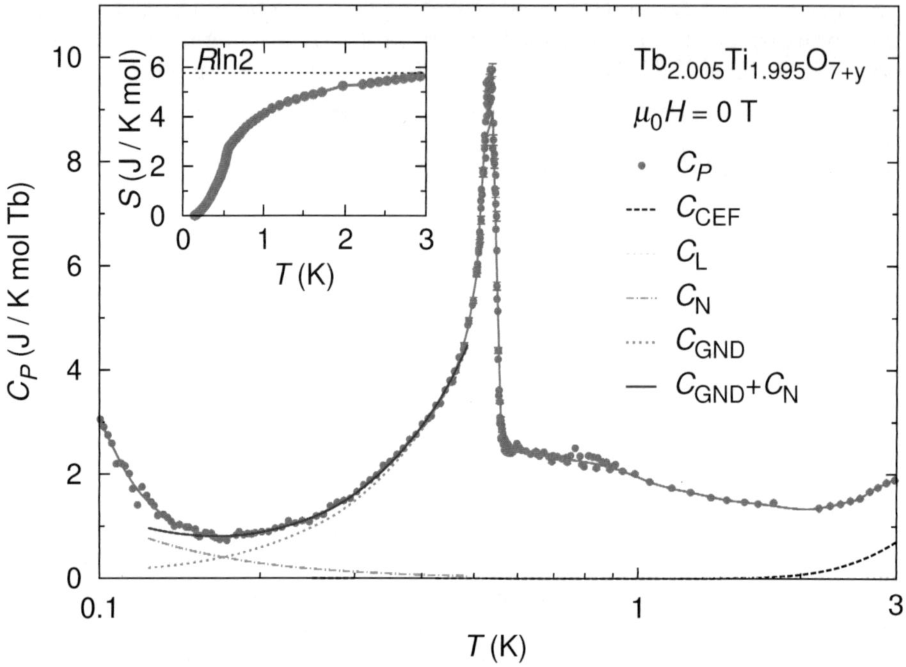


**Figure 1.** Temperature dependence of the specific heat $C_P$ of single-crystalline $Tb_{2+x}Ti_{2-x}O_{7+y}$ with $x = 0.005$ at zero field. The red filled circles represent the measured $C_P$ data from 3 to 0.1 K. $C_{CEF}$, $C_L$, $C_N$, and $C_{GND}$ represent the calculated or fitted results of the contribution from excited CEF to the specific heat, the lattice specific heat, the nuclear specific heat, and the specific heat of the low-$T$ excitations of the phase transition, respectively. Inset shows the temperature dependence of the entropy evaluated from the integration of $(C_P - C_{CEF} - C_L - C_N)/T$ from 0.15 K.

0.15 K. Although we extended to fit the data using higher order terms of $C_N$ obtained by both considerations of magnetic dipole hyperfine coupling and electric nuclei-quadrupole coupling [36], the fitting was not improved. This may reflect low energy fluctuations or other anomalous contributions such as photon-like excitations and the proximity of quantum criticality [5, 37]. It is also considered that a coupling between low energy fluctuations of $4f$ moments and nuclear spins (and also quadrupole moments) of the Tb nuclei may give rise to the anomalous enhancement of $C_P$. Indeed, the energy scale of 0.1 K corresponds to 10 $\mu$eV and then the inelastic neutron spectrum previously observed around $E = 0$ [5] could be ascribable to the possible existence of such low energy excitations, although higher resolution experiments are needed to clarify this point. Otherwise, the anomalous enhancement of $C_P$ might be related to the lowering of the thermal conductivity of the sample on cooling, which may cause temperature gradient inside the sample and overestimation of the $C_P$ value, although the small and thin crystal was used for the experiments. Note that such a large enhancement of $C_P$ has been also observed in a QSL of the metallic pyrochlore $Pr_2Ir_2O_7$ [38]. More precise and careful measurements are required for further understanding. The fitting yields $A = 24(2)$, $n = 2.4(1)$, and $B = 0.014(2)$. Since the coefficient $B$ is written to be $B = R(1.25\alpha'^2 + P^2)$ using the magnetic hyperfine constant $\alpha'$, quadrupole coupling constant $P$, and gas constant $R$ [36], it is roughly estimated that $\alpha' \sim 0.037$ K when $P = 0$ or $P \sim 0.041$ K when $\alpha' = 0$ (or then $\alpha'$ and $P$ are expected within half values of those). These values are the same order of the values for systems including Tb

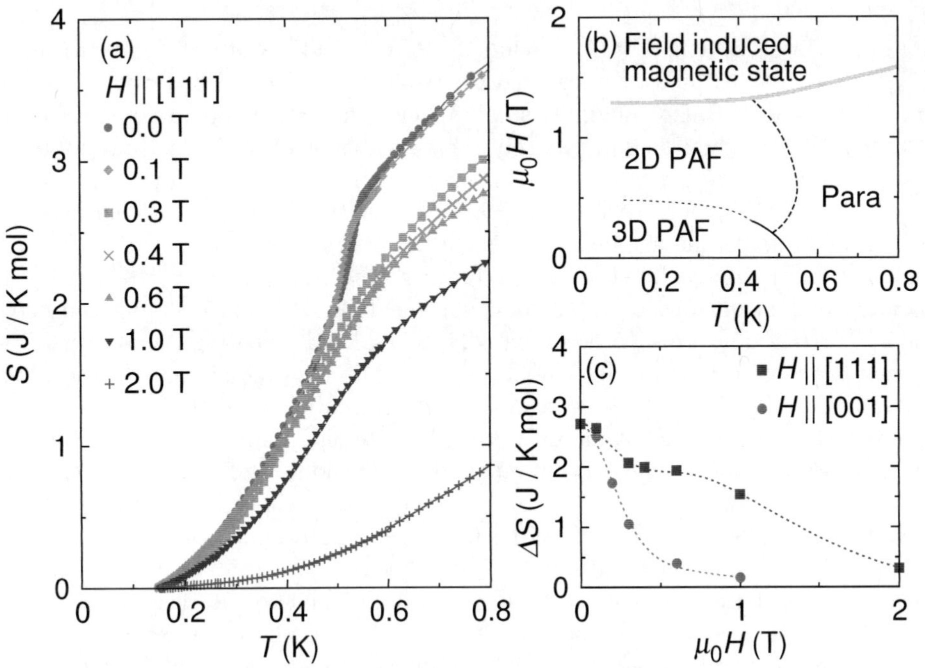


**Figure 2.** (a)Temperature dependence of the entropy for several magnetic fields for $H||[111]$. (b)$H$–$T$ phase diagram for $H||[111]$ [20]. The black solid line up to 0.4 T indicates the phase boundary between the 3D quadrupolar state of the planar antiferropseudospin (PAF), 3D PAF, and the paramagnetic paraquadrupole state. The black dashed line is the phase boundary between the 2D quadrupolar state, 2D PAF, and the paramagnetic paraquadrupole state. The black dotted line represents the line of the crossover or phase transition between the 3D and 2D PAF states. The gray solid line is the putative phase boundary between the 2D PAF state and the field-induced magnetic state, which is estimated from a kink of the magnetization. (c)Field dependence of the entropy change $\Delta S[= S(0.55\text{K}, H) - S(0.15\text{K}, H)]$ for $H||[111]$ and $H||[001]$. The lines are guides to the eyes.

nuclei and the theoretical expectation for the case of the Tb metal [39, 40]. We thus considered that the estimation of $C_N$ is approximately reasonable and intended to focus on the data above 0.15 K at present for the evaluation of the entropy, which was calculated from the integration of $\Delta C/T = (C_P - C_{\text{CEF}} - C_{\text{L}} - C_{\text{N}})/T$. The corresponding entropy change in temperature is shown in the inset of Fig. 1.

It is remarkable that the entropy in zero field at $T_c$ is 2.7 J K$^{-1}$mol$^{-1}$. This value is about 50% of $R \ln 2$ expected for a non-Kramers doublet of the Tb ions. Even when we consider the low-$T$ contribution below 0.15 K, it doesn't reach $R \ln 2$. Instead, the entropy release persists to higher temperatures and is saturated around 3 K, as seen in the inset of Fig. 1. This behavior is similar to that of previous experiments for a sample showing LRO [41]. These results imply that strong fluctuations still persist to temperatures higher than $T_c$. It is also suggested that a QSL-like state or the proximity of it could realize at temperatures between near $T_c$ and $\sim 1$ K, where $C_P$ slightly increases like a Schottky anomaly signified by SI correlations and the value is also large enough ($\sim 2$ J K$^{-1}$mol-Tb$^{-1}$): these imply finite excitation density of states or large magnetic fluctuations of quantum or thermally-excited monopoles, although $S$ does not exhibit the plateau of the residual entropy, which is masked or must be released by the sharp peak of

$C_P$ at $T_c$.

Figure 2(a) shows the temperature dependence of the field variation of the entropy, $S(T,H)$, for $H||[111]$. For these estimations, we used the same method described above, assuming that $C_{CEF}$ and $C_L$ in fields are also negligibly small at low temperatures below 1 K. Interestingly, it is found that $S(T,H)$ exhibits quantitatively the same behavior in fields around 0.5 T, while it decreases (or increases) with increasing (or decreasing) magnetic fields in a temperature range up to 0.8 K. These behaviors reflect the properties of the $H$–$T$ phase diagram for $H||[111]$ [Fig. 2(b)], where the 3D quadrupolar state is considered to be replaced by the 2D quadrupolar state, and then by the magnetic state at low temperatures. In fact, as plotted in Fig. 2(c), the field dependence of the entropy change around zero-field $T_c$, $\Delta S$, exhibits a plateau-like behavior around 0.5 T for $H||[111]$. This decreases rapidly for $H||[001]$ reflecting the relatively small field that induces the magnetic state above 0.3 T [42]. It is considered that at the intermediate field ($\sim$ 0.5 T) for $H||[111]$, the phase transition occurs only due to the change of the state in each kagomé-lattice layer, which could only contribute to the change in entropy. Therefore, this situation may lead to almost the same temperature dependence of $S(T,H)$ around 0.5 T and the plateau state of $\Delta S$.

Now we find that the difference of $\Delta S$ between at 0 and around 0.5 T for $H||[111]$ is about 0.7 J $K^{-1}mol^{-1}$. Then, a question is what does this value mean? An interesting scenario is that, if the states in 0 and about 0.5 T around 1 K are similar to the classical SI and KI states and the residual entropies of these states are released by the phase transition and then zero as expected from the quantum model [43], the difference of $\Delta S$ between at 0 and around 0.5 T may possibly correspond to the difference between the SI residual entropy (1.68 J $K^{-1}mol^{-1}$) [44] and the KI residual entropy (0.67 J $K^{-1}mol^{-1}$) [45, 46]; i.e., $\sim$ 1 J $K^{-1}mol^{-1}$. In fact, the value of 0.7 J $K^{-1}mol^{-1}$ is close to this value. It is considered that this entropy will be released at temperatures higher than 0.8 K. For this context, further detailed experiments at higher temperatures under magnetic field are interesting future subjects.

## 4. Conclusion

In conclusion, we reported thermodynamic properties of the quadrupole-order sample of $Tb_{2+x}Ti_{2-x}O_{7+y}$ with $x = 0.005$. We observed that the specific heat shows a minimum around 2 K and slightly increases on cooling, which is similar to a Schottky-type anomaly for canonical spin ices. A clear peak at $T_c = 0.53$ K results in the phase transition to the quadrupolar state. The entropy change at $T_c$ is 2.7 J $K^{-1}mol^{-1}$, which is smaller than $R\ln 2$ and suggests the strong fluctuations that remains at higher temperatures than $T_c$. The field dependence of the entropy change around 0.5 K and in 0.5 T for $H||[111]$ exhibits a plateau. This characteristic feature probably reflects the formation of the two-dimensional quadrupolar state by the [111] magnetic field.

## Acknowledgments

We thank S. Onoda, Y. Kato, R. Higashinaka, M. Wakita, and H. Kageyama for useful discussions and their assistance. This work was supported by JSPS KAKENHI grant numbers 25400345 and 26400336. One of the specific heat measurements was performed using facilities of ISSP, University of Tokyo. One of the authors would like to acknowledge the support from the Motizuki Fund of Yukawa Memorial Foundation.

## References

[1] Lacroix C, Mendels P and Mila F (eds) 2011 *Introduction to Frustrated Magnetism* (Berlin, Heidelberg: Springer)
[2] Gingras M J P and McClarty P A 2014 *Rep. Prog. Phys.* **77** 056501
[3] Gardner J S, Dunsiger S R, Gaulin B D, Gingras M J P, Greedan J E, Kiefl R F, Lumsden M D, MacFarlane W A, Raju N P, Sonier J E, Swainson I and Tun Z 1999 *Phys. Rev. Lett.* **82** 1012

[4] Gardner J S, Keren A, Ehlers G, Stock C, Segal E, Roper J M, Fak B, Stone M B, Hammar P R, Reich D H and Gaulin B D 2003 *Phys. Rev. B* **68** 180401
[5] Taniguchi T, Kadowaki H, Takatsu H, Fak B, Ollivier J, Yamazaki T, Sato T J, Yoshizawa H, Shimura Y, Sakakibara T, Hong T, Goto K, Yaraskavitch L R and Kycia J B 2013 *Phys. Rev. B* **87** 060408(R)
[6] Wakita M, Taniguchi T, Edamoto H, Takatsu H and Kadowaki H 2016 *J. Phys.: Conf. Ser.* **683** 012023
[7] Gingras M J P, den Hertog B C, Faucher M, Gardner J S, Dunsiger S R, Chang L J, Gaulin B D, Raju N P and Greedan J E 2000 *Phys. Rev. B* **62** 6496
[8] Kao Y J, Enjalran M, Maestro A D, Molavian H R and Gingras M J P 2003 *Phys. Rev. B* **68** 172407
[9] Molavian H R, Gingras M J P and Canals B 2007 *Phys. Rev. Lett.* **98** 157204
[10] Bonville P, Mirebeau I, Gukasov A, Petit S and Robert J 2011 *Phys. Rev. B* **84** 184409
[11] Petit S, Guitteny S, Robert J, Bonville P, Decorse C, Ollivier J, Mutka H and Mirebeau I 2015 *EPJ Web Conf.* **83** 03012
[12] Guitteny S, Mirebeau I, de Réotier P D, Colin C V, Bonville P, Porcher F, Grenier B, Decorse C and Petit S 2015 *Phys. Rev. B* **92** 144412
[13] Fritsch K, Ross K A, Qiu Y, Copley J R D, Guidi T, Bewley R I, Dabkowska H A and Gaulin B D 2013 *Phys. Rev. B* **87** 094410
[14] Fritsch K, Kermarrec E, Ross K A, Qiu Y, Copley J R D, Pomaranski D, Kycia J B, Dabkowska H A and Gaulin B D 2014 *Phys. Rev. B* **90** 014429
[15] Kermarrec E, Maharaj D D, Gaudet J, Fritsch K, Pomaranski D, Kycia J B, Qiu Y, Copley J R D, Couchman M, Morningstar A, Dabkowska H A and Gaulin B D 2015 *Phys. Rev. B* **92** 245114
[16] Fennell T, Kenzelmann M, Roessli B, Mutka H, Ollivier J, Ruminy M, Stuhr U, Zaharko O, Bovo L, Cervellino A, Haas M K and Cava R J 2014 *Phys. Rev. Lett.* **112** 017203
[17] Ruminy M, Bovo L, Pomjakushina E, Haas M K, Stuhr U, Cervellino A, Cava R J, Kenzelmann M and Fennell T 2016 *Phys. Rev. B* **93** 144407
[18] Ruminy M, Groitl F, Keller T and Fennell T arXiv:1607.07688
[19] Kadowaki H, Takatsu H, Taniguchi T, Fåk B and Ollivier J 2015 *SPIN* **5** 1540003
[20] Takatsu H, Onoda S, Kittaka S, Kasahara A, Kono Y, Sakakibara T, Kato Y, Fåk B, Lynn J O J W, Taniguchi T, Wakita M and Kadowaki H 2016 *Phys. Rev. Lett.* **116** 217201
[21] Onoda S and Tanaka Y 2010 *Phys. Rev. Lett.* **105** 047201
[22] Onoda S and Tanaka Y 2011 *Phys. Rev. B* **83** 094411
[23] Matsuhira K, Hiroi Z, Tayama T, Takagi S and Sakakibara T 2002 *J. Phys.: Condens. Matter* **14** L559
[24] Hiroi Z, Matsuhira K, Takagi S, Tayama T and Sakakibara T 2002 *J. Phys. Soc. Jpn.* **72** 411
[25] Higashinaka R, Fukazawa H and Maeno Y 2003 *Phys. Rev. B* **68** 014415
[26] Sakakibara T, Tayama T, Hiroi Z, Matsuhira K and Takagi S 2003 *Phys. Rev. Lett.* **90** 207205
[27] Tabata Y, Kadowaki H, Matsuhira K, Hiroi Z, Aso N, Ressouche E and Fåk B 2006 *Phys. Rev. Lett.* **97** 257205
[28] Takatsu H, Goto K, Otsuka H, Higashinaka R, Matsubayashi K, Uwatoko Y and Kadowaki H 2013 *J. Phys. Soc. Jpn.* **82** 073707
[29] Otsuka H, Takatsu H, Goto K and Kadowaki H 2014 *Phys. Rev. B* **90** 144428
[30] Harris M J, Bramwell S T, Holdsworth P C W and Champion J D M 1998 *Phys. Rev. Lett.* **81** 4496
[31] Mirebeau I, Bonville P and Hennion M 2007 *Phys. Rev. B* **76** 184436
[32] Ruminy M, Valdez M N, Wehinger B, Bosak A, Adroja D T, Stuhr U, Iida K, Kamazawa K, Pomjakushina E, Prabakharan D, Haas M K, Bovo L, Sheptyakov D, Cervellino A, Cava R J, Kenzelmann M, Spaldin N A and Fennell T 2016 *Phys. Rev. B* **93** 214308
[33] Elliott R J and Stevens K W H 1953 *Proc. Roy. Soc. A* **218** 553
[34] Bleaney B and Hill R W 1961 *Proc. Roy. Soc.* **78** 313
[35] Gopal E S R 1966 *Specific Heats at Low Temperatures* (New York: Plenum Press)
[36] Lounasmaa O V and Roach P R 1962 *Phys. Rev.* **128** 622
[37] Hermele M, Fisher M P A and Balents L 2004 *Phys. Rev. B* **69** 064404
[38] Tokiwa Y, Ishikawa J J, Nakatsuji S and Gegenwart P 2014 *Nature Mater.* **13** 356
[39] Kondo J 1961 *J. Phys. Soc. Jpn.* **16** 1690
[40] Bleaney B 1963 *J. Appl. Phys.* **16** 1690
[41] Hamaguchi N, Matsushita T, Wada N, Yasui Y and Sato M 2004 *Phys. Rev. B* **69** 132413
[42] Takatsu H, Taniguchi T, Kittaka S, Sakakibara T and Kadowaki H 2016 *J. Phys.: Conf. Ser.* **683** 012023
[43] Kato Y and Onoda S 2015 *Phys. Rev. Lett.* **115** 077202
[44] Ramirez A P, Hayashi A, Cava R J, Siddharthan R and Shastry B S 1999 *Nature* **399** 333
[45] Udagawa M, Ogata M and Hiroi Z 2002 *J. Phys. Soc. Jpn.* **71** 2365
[46] Moessner R and Sondhi S L 2003 *Phys. Rev. B* **68** 184512

# Low temperature magnetic properties of frustrated pyrochlore ferromagnet $Ho_2Ir_2O_7$

**Dinesh Kumar[1], S Y Chen[1], M K Lee[1], C M N Kumar[2,3], R Aldus[2] and L J Chang[1]**

[1] Department of Physics, National Cheng Kung University, No. 1, University Road, Tainan 70101, Taiwan

[2] Jülich Centre for Neutron Science JCNS, Forschungszentrum Jülich GmbH, Outstation at SNS, Oak Ridge National Laboratory, Oak Ridge, Tennessee 37831, United States

[3] Laboratoire National des Champs Magnetiques Intenses CNRS, 143, avenue de Rangueil, F-31400 Toulouse, France

E-mail: dinidixit[at]gmail.com

**Abstract.** We have performed low temperature magnetic susceptibility, specific heat, electrical resistivity and neutron diffraction measurements to study the magnetic properties of frustrated pyrochlore ferromagnet $Ho_2Ir_2O_7$. The magnetization and dc resistivity measurement indicate metal-insulator transition near 140 K. The accumulated value of observed entropy calculated from heat capacity data is $R\ln 2$, indicating the $Ho^{3+}$ doublet ground state. The Ir sublattice order is antiferromagnetic which is evident from the magnetization measurements. The best fit of the neutron diffraction spectra at 1.8 K shows the magnetic structures with two types of components. One of the magnetic structure is all-in-all-out type, which is due to the strong coupling between $Ho^{3+}$ and $Ir^{4+}$ ions and develops below 18 K. It suggests that the magnetic moment of $Ho^{3+}$ ions align in the field of $Ir^{4+}$, ordering along local <111> directions. There is also a $Ho^{3+}$ magnetic component perpendicular to <111> below 100 K. This perpendicular component may originate from the defects in the Ir sub-lattice. Defects such as Ir deficiency or an Ir atom pointing at the opposite direction will cause the emergent of the perpendicular component.

## 1. Introduction

The study of frustrated pyrochlore oxides having the composition $A_2B_2O_7$ is very crucial to understand the ground state that results from the competing interactions of the spins on the corners of the tetrahedron lattice. In case of the pyrochlore ferromagnet, the local Ising anisotropy makes the spins to point either into or out of the center of tetrahedron and consequently the spin configurations on ground state of one tetrahedron are six-fold degenerate having two-in-two-out structures. The pyrochlore lattice consists of a three dimensional network of corner sharing tetrahedra. If we extend this two-in-two-out structure to the entire pyrochlore lattice, we obtain a macroscopic ground state, which eventually lead to the residual entropy. The macroscopic degenerate ground state, a feature possessed by a number of frustrated materials, has been a fascinating area of research. Frustrated magnetism on site A leads to one of the remarkable phenomenon termed as "spin ice" behaviour where residual entropy has been observed in

pyrochlore materials e.g. $Ho_2Ti_2O_7$ [1], $Dy_2Ti_2O_7$ [2], $Ho_2Sn_2O_7$ [3]. For the pyrochlore lattice, it is a well-established fact that antiferromagnetic interactions lead to the high degree of geometrical frustration [4]. Interestingly it is also been observed that even in case of ferromagnetic interaction in pyrochlore lattice, the frustration may arise due to the strong single-ion anisotropy [5].
The iridium based pyrochlore compounds, $R_2Ir_2O_7$ (R = Pr, Nd, Sm and Eu) have been investigated and a metal-nonmetal changeover has been observed in these compounds [6]. The $Ir^{4+}$ ions containing five *5d* electrons differ from identical $Mo^{4+}$ and $Ru^{4+}$ ions as follows: $Ir^{4+}$ has low spin $S = ½$ configuration, while $Mo^{4+}$ and $Ru^{4+}$ have $S = 1$ configuration and consequently $Ir^{4+}$ brings about different properties. In this paper, we present the low temperature magnetic properties of frustrated pyrochlore $Ho_2Ir_2O_7$. The magnetization measurements indicate metal-insulator transition near 140 K. Neutron diffraction experiments reveal two types of the components of $Ho^{3+}$ in the magnetic structure of pyrochlore $Ho_2Ir_2O_7$.

## 2. Experimental details

Polycrystalline samples of $Ho_2Ir_2O_7$ have been prepared using conventional solid-state synthesis technique. The starting materials were $Ho_2O_3$ (99.99%) powder and $IrO_2$ (99.99%) powder in the ratio 1:2 respectively. The materials were grinded well with a mortar and pestle and we pre-sintered the samples at $800^0$ C for 12 hours. After pre-sintering of the samples, we pelletized the samples and heated at $1100^0$ C for 20 days. The phase purity of samples is confirmed with synchrotron X-ray diffraction measurements. The dc magnetization is performed with a superconducting quantum interference device (SQUID) magnetometer (Quantum Design, USA) for various temperatures down to 2 K and magnetic fields up to 5 T. The specific heat measurement is performed using a relaxation method in a physical property measurement system (PPMS) (Quantum Design, USA) down to 1.8 K. The AC susceptibility measurements has been performed with PPMS with $H_{ac}$ = 10 Oe superimposed on various dc magnetic fields up to 2 T, in the temperature range of 2 K - 30 K. Neutron powder diffraction measurements were carried out on the time-of-flight powder diffractometer POWGEN at the Spallation Neutron Source, Oak Ridge National Laboratory. Diffraction data were collected with the center wavelength 1.066 Å. Rietveld refinements for structures of $Ho_2Ir_2O_7$ were performed using the FullProf program.

## 3. DC magnetization measurements

Temperature dependence of the dc susceptibility of $Ho_2Ir_2O_7$ at 50 Oe and 1 kOe in the temperature range 50 K – 300 K is shown in figure 1. The ZFC and FC curves at 1 kOe are observed to be separated around 141 K. This ZFC-FC behaviour is consistent with the earlier reported studies on the family of these compounds, indicating that it shows metal-insulator transition (MIT) [7]. The shape of FC curve below $T_{MI}$ depends upon the preparation of the sample. In our case, below $T_{MI}$, FC lies above the ZFC curve for both field values. (Note that FC and ZFC for H = 1 kOe have been divided by a factor of 1.65 for scaling purposes). One interesting feature of the FC curve at 1 kOe is that it starts separating apart ZFC curve around 225 K, 84 K higher than the $T_{MI}$. Previous studies have reported similar effects for metallic pyrochlore compound $Pr_2Ir_2O_7$, where resistivity starts to increase below $T_{MI}$ [8]. This effect is attributed to the enhanced magnetic scattering of the charge carriers while approaching the magnetic ordering, due to spin-ice behaviour in a frustrated pyrochlore. The temperature dependence of inverse susceptibility (not shown here) and the Curie Weiss law fitting of H – 50 Oe data give magnetic moment of $Ho^{3+}$ = 8.362(1) $\mu_B$ and $\theta_w$ = 1.9 K, indicating the ferromagnetic coupling of spins in $Ho_2Ir_2O_7$.

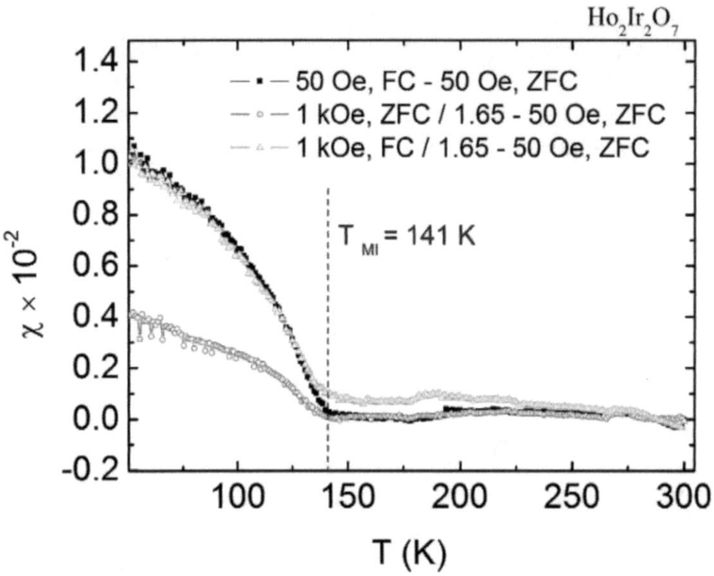


Figure 1: Temperature dependence of the susceptibility of $Ho_2Ir_2O_7$ at H = 50 Oe and 1 kOe. The data for 1 kOe is divided by a factor 1.65 for data scaling. The dotted blue line represents metal insulator transition temperature ($T_{MI}$) ~ 141 K. The data for 50 Oe (ZFC) is subtracted from each data set.

## 4. AC Susceptibility Measurements

Figures 2(a) and 2(b) show the temperature dependence of the real (χ') and imaginary (χ'') parts of the ac susceptibility of $Ho_2Ir_2O_7$ for $H_{dc}$ = 0.7 T and $H_{ac}$ = 10 Oe and under different ac field frequencies ranging between 10 Hz – 10 kHz. As shown in figure 2(a), χ' shows a peak near 5 K, which is consistent with the earlier published results of spin-ice materials and this peak is associated with the dynamic spin freezing at the low temperature [3,9,10]. From figure 2(b) we observed that χ'' shows a broad peak around 6 K, which is also associated with freezing of spins at the low temperature. The maxima in χ' and χ'' decreases with the increase in the ac frequency which is expected for spin freezing.

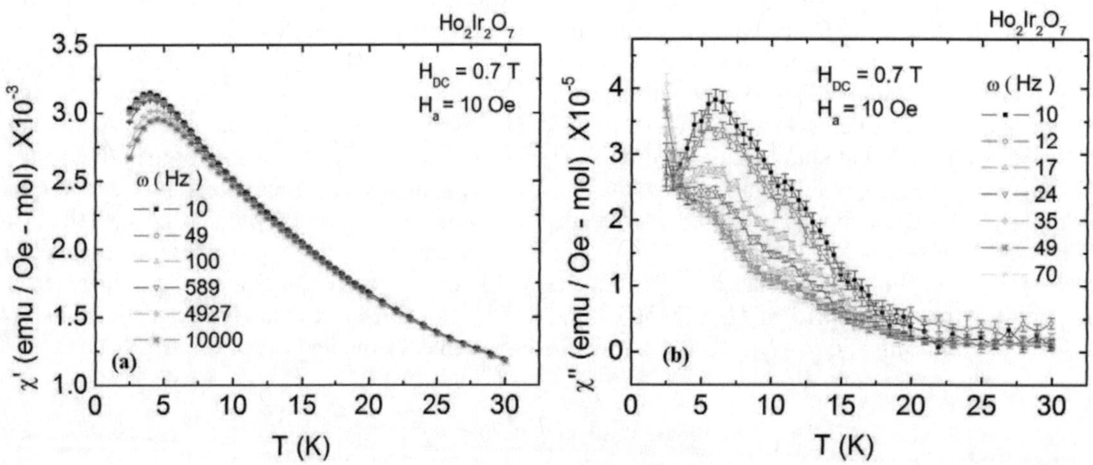


Figure 2: (a) Temperature variation of the real part of the ac susceptibility (χ') of $Ho_2Ir_2O_7$ at various measurement frequencies ranging between 10 Hz – 10 kHz with $H_{dc}$ = 0.7 T and $H_{ac}$ = 10 Oe. (b) Temperature variation of the imaginary part of the ac susceptibility (χ'') of $Ho_2Ir_2O_7$ at various measurement frequencies ranging between 10 Hz – 70 Hz with $H_{dc}$ = 0.7 T and $H_{ac}$ = 10 Oe.

IOP Conf. Series: Journal of Physics: Conf. Series **828** (2017) 012008 doi:10.1088/1742-6596/828/1/012008

The monotonic decrease in the in $\chi$' and $\chi$'' with increase in the temperature is in well agreement with the published results for the spin-ice materials [3]. We also observed that peak in $\chi$'' shifts to the lower temperature with increase in the ac frequency which indicates the relaxation of the magnetization in $Ho_2Ir_2O_7$ at low temperatures.

## 5. Heat Capacity measurements

The magnetic contribution to the specific heat ($C_m$) is obtained by subtracting lattice contribution (specific heat of pyrochlore $Lu_2Sn_2O_7$), and nucleon hyperfine contribution, which has been estimated from Schottky specific heat, from the total specific heat ($C_p$) of $Ho_2Ir_2O_7$ in zero field. The magnetic entropy is calculated by integrating $C_m/T$ over temperature using the formula: $\Delta S_m = \int_{T'}^{T''} \frac{C}{T}\, dT$, where T' and T'' are the initial and final temperature taken for integration interval. As we can observe from the temperature dependence of entropy (as shown in figure 3), zero-point entropy of R/2 ln3/2 does not appear in $Ho_2Ir_2O_7$ and $\Delta S_m$ is saturated near 30 K and attains the value 5.92 J/mol-K, which is slightly more than spin freezing condition Rln2 = 5.76 J/mol-K. Similar results are reported for $Dy_2Ir_2O_7$ [6]. In case of $Dy_2Ir_2O_7$ the spin freezing condition is obtained near 20 K.

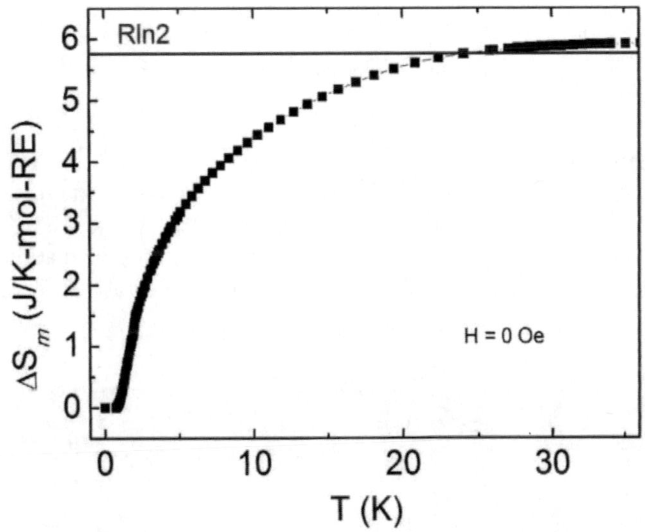


Figure 3: The temperature variation of the magnetic entropy at H = 0 Oe in the temperature range 1.8 K – 45 K.

## 6. Neutron diffraction measurements

Neutron powder diffraction experiments had been carried out on $Ho_2Ir_2O_7$ at various temperatures (the detailed results to be published elsewhere). Figure 4 shows the Reitveld refinement data on $Ho_2Ir_2O_7$ at T = 29 K. The red dots represent the experimental data. The solid black line is the theoretically calculated intensity. The peaks (420) and (222) might be possibly appear because of the lattice defects and not from the magnetic structure. The blue curve represents the difference of both values. The best fit of the neutron diffraction spectra at 1.8 K shows the magnetic structures with two types of components. One of the magnetic structure is all-in-all-out type, which is due to the strong coupling between $Ho^{3+}$ and $Ir^{4+}$ ions and develops below 18 K. It suggests the magnetic moment of $Ho^{3+}$ ions align in the field of $Ir^{4+}$ ordering along local <111> directions [11,12]. Another magnetic component of $Ho^{3+}$, perpendicular to <111> is observed below 100 K. This perpendicular component may originate from the defects in the Ir sublattice. Defects such as Ir deficiency or an Ir atom pointing at the opposite direction will cause the emergent of the perpendicular component [11].

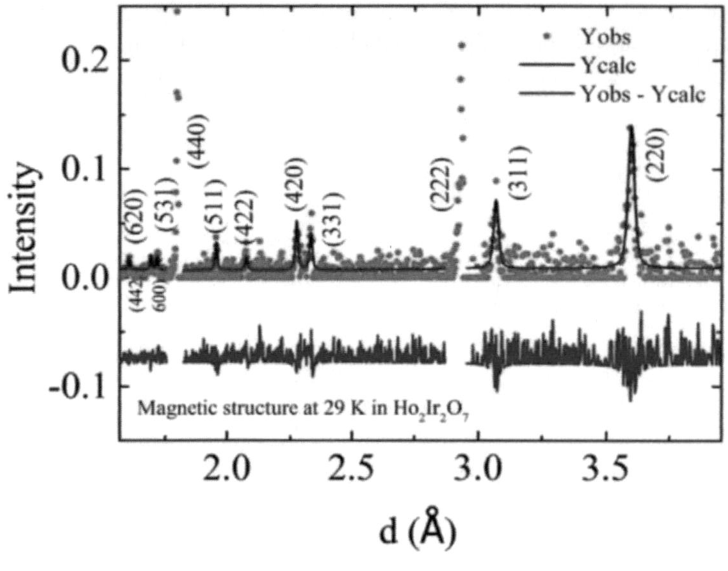


Figure 4: Reitveld refined neutron diffraction data of $Ho_2Ir_2O_7$ at T = 29 K. The red dots represent the observed intensity; black curve represent the theoretically calculated intensity and the blue curve represent the difference of both.

## 7. Resistivity measurements

We have performed resistivity measurements with standard four probe method. The temperature variation of the resistivity is shown in figure 5(a) in the temperature range 1.85 K – 7 K and under various magnetic field values ranging up to 5 T. The sample shows insulating behavior at low temperature (1.85 K) and H = 0 Oe. We have observed that with the increase in the field, the electrical

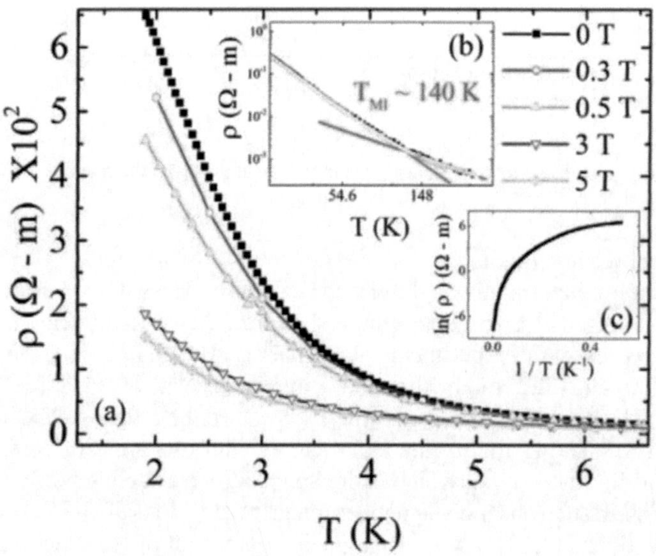


Figure 5: (a) Temperature variation of the electrical resistivity at various magnetic fields ranging up to 5 T. (b) Temperature variation of the resistivity at high temperatures indicating the metal-insulator transition near 140 K. (c) 1/T variation of the ln(ρ) to analyse thermistor behaviour of $Ho_2Ir_2O_7$.

behavior of the $Ho_2Ir_2O_7$ sample changes from insulator to the metallic behavior. Figure 5(b) shows the temperature variation of the resistivity at the high temperature and indicates the metal-insulator transition near 140 K. Further we have analyzed the thermistor behavior of $Ho_2Ir_2O_7$ as shown in the figure 5(c) and find that 1/T variation of ln(ρ) does not follow the common semiconductor thermistors formula $1/T = A + B \ln(\rho) + C [\ln(\rho)]^3$ where A, B and C are constants [13].

## 8. Conclusion

In conclusion, we have performed low temperature ac and dc susceptibility, specific heat, electrical resistivity and neutron diffraction measurements to study the magnetic properties of frustrated pyrochlore ferromagnet $Ho_2Ir_2O_7$. The dc magnetization and resistivity measurements show metal-insulator transition near 140 K. The total entropy accumulated from heat capacity data is Rln2, which indicates the $Ho^{3+}$ doublet ground state. The Ir sublattice orders is the antiferromagnetic which is evidenced by the magnetization measurements. The magnetic structures with two types of components are present in the best fit of the neutron diffraction spectra at 1.8 K. One of the magnetic structure component is all-in-all-out configuration of Ising spins, along local <111> axes and develops below 18 K. There is also a component of magnetic structure, which is perpendicular to <111> axes. The perpendicular component of the magnetic structure may possibly originate from the defects in the Ir sub-lattice.

## Acknowledgements

The authors would like to acknowledge financial support provided by Ministry of Science and Technology (MOST), Taiwan under project numbers 'MOST 104-2112-M-006-010-MY3' and 'MOST 104-2119-M-006-017-MY3'. The authors also gratefully acknowledge the financial support provided by JCNS to perform the neutron scattering measurements at the Spallation Neutron Source (SNS), Oak Ridge, USA. Part of the research conducted at SNS was sponsored by the Scientific User Facilities Division, Office of Basic Energy Sciences, US Department of Energy.

## References

[1] Harris M J, Bramwell S T, McMorrow D F, Zeiske T and Godfrey K W 1997 *Phys. Rev. Lett.* **79** 2554
[2] Ramirez A P, Hayashi A, Cava R J, Siddharthan R and Shastry*et* B S 1999 *Nature* **399** 333
[3] Matsuhira K, Hinatsu Y, Tenya K and Sakakibara T 2000 *J. Phys.: Condens. Matter* **12** L649
[4] Harris M J and Zinkin M P 1996 *Mod. Phys. Lett. B* **10** 417
[5] Harris M J, Bramwell S T, Zeiske T, McMorrow D F, King P J C 1998 *J. Magn. Mag. Materials* **177** 757
[6] Yanagishima D and Maeno Y 2001 *J. Phys. Soc. Japan* **70** 2880
[7] Matsuhira K, Wakeshima M, Hinatsu Y and Takagi S 2011 *J. Phys. Soc. Japan* **80** 094701
[8] Nakatsuji S, Machida Y, Maeno Y, Tayama T, Sakakibara T, Duijn J van, Balicas L, Millican J N, Macaluso R T and Chan J Y 2006 *Phy. Rev. Lett.* **96** 087204
[9] Matsuhira K, Y Hinatsu Y and Sakakibara T 2001 *J. Phys.: Condens. Matter* **13** L737
[10] Lau G C, Freitas R S, Ueland B G, Muegge B D, Duncan E L, Schiffer P and Cava R J 2006 *Nature Physics* **2** 249-253
[11] Lefrançois E, Simonet V, Ballou R, Lhotel E, Hadj-Azzem A, Kodjikian S, Lejay P, Manuel P, Khalyavin D and Chapon L C 2015 *Phys. Rev. Lett.* **114** 247202
[12] Tomiyasu K, Matsuhira K, Iwasa K, Watahiki M, Takagi S, Wakeshima M, Hinatsu Y, Yokoyama M, Ohoyama K, and Yamada K 2012 *J. Phys. Soc. Japan* **81** 034709
[13] Steinhart J S and Hart S R 1968 Deep Sea Research and Oceanographic Abstracts **15** 497-503

IOP Conf. Series: Journal of Physics: Conf. Series **828** (2017) 012009 doi:10.1088/1742-6596/828/1/012009

# Evolution of magnetic and transport properties in hole doped $Y_2Ir_2O_7$

**Harish Kumar and A. K. Pramanik**
School of Physical Sciences, Jawaharlal Nehru University, New Delhi - 110067, India.
E-mail: harishjurel@gmail.com

**Abstract.** We report the effect of $Ru^{4+}$ substitution at Ir-site on structural, magnetic and electrical properties in pyrochlore iridate $Y_2Ir_2O_7$. Structural investigation has been done using x-ray powder diffraction and Rietveld analysis for both samples. There is no structural phase transition with Ru doping. Magnetization study shows that on slightly substitution of Ru at Ir-site effectively shifts the magnetic irreversible temperature $T_{irr}$ towards low temperature and reduce the bifurcation between $M_{ZFC}$ and $M_{FC}$ branches. The electrical resistivity shows an insulating behavior in whole temperature range, however, resistivity decreases with doping of Ru.

## 1. Introduction

Pyrochlore iridates $A_2B_2O_7$ (A = rare earth elements) are very interesting showing exotic electronic and magnetic properties [1]. Within the extended $5d$ character, Ir exhibit reduced electronic correlation ($U$), however, the heavy nature of Ir gives prominent spin orbit coupling (SOC) effect. Moreover, the interpenetrating layers of $IrO_6$ octahedra and $OA_4$ tetrahedra introduce geometrical frustration. This frustration is seen to give rise many interesting phenomena such as, spin glass, spin ice, spin liquid, etc. The physical properties interestingly evolve with A-site element: from magnetic insulating to complex metallic phases [2]. When the $A^{3+}$ ion is magnetic, there is a possibility of $f - d$ exchange interactions between $Ir^{4+}$ and $A^{3+}$ ions which naturally introduces complicated magnetic interactions in Ir-based pyrochlores [3]. The $Y_2Ir_2O_7$ is of special interest as the nonmagnetic $Y^{3+}$ resides at A-site. In this sense, $Ir^{4+}$ ion ($5d^5$) with single unpaired electron generate frustration in this material. The $Y_2Ir_2O_7$ shows an insulating behavior where the resistivity increases with decreasing temperature. $Y_2Ir_2O_7$ shows irreversibility in magnetization data around 160 K [4, 5, 6]. Given that frustration plays vital role in pyrochlore lattice the spin freezing to glassy state could also be a probable magnetic state in $Y_2Ir_2O_7$ at low temperature [4, 7, 8].

In this work, we have studied the effect of $Ru^{4+}$ substitution in pyrochlore iridate $Y_2Ir_2O_7$. The $5d$ based Ir has strong SOC and less $U$, on other hand, $4d$ Ru has more $U$ and less SOC. Thus, this substitution is expected to reduce the SOC and intensify the electron correlation which will result in change in ground state properties. Substitution of $Ru^{4+}$ ($4d^4$) at $Ir^{4+}$ ($5d^5$) site not only reduce the SOC but also introduce a hole in $t_{2g}$ band, which reduce the level of Fermi surfaces $E_F$, thus, moving the system away from the insulating to metallic state. We have studied structural, magnetic and electrical transport properties. we find that magnetic moment, Curie temperature, frustration parameters ($f$) and magnetic irreversible temperature

$T_{irr}$ decreases. Moreover, Resistivity data of $Y_2Ir_2O_7$ show insulating behavior throughout the temperature range, however, on Ru substitution, resistivity is decreased.

## 2. Experimental Procedure

Polycrystalline samples of $Y_2Ir_{2x}Ru_xO_7$ (x = 0.0 and 0.1) have been prepared using standard solid state method. The mixture of ingredient powder materials $Y_2O_3$ , $RuO_2$ and $IrO_2$ with phase purity 99.99% (Sigma-Aldrich) taken in stoichiometric ratio are ground well. Pellet is heated in air in temperature range 1000 $^oC$ to 1160 $^oC$ for 18 days with several intermediated grindings. The material is characterized by powder x-ray diffraction (XRD) measured with Rigaku-made Miniflex600 and PANalytical XPert powder diffractometer. The XRD pattern is refined with Rietveld program. DC Magnetization ($\chi$) data have been collected using a vibrating sample magnetometer (PPMS, Quantum Design).

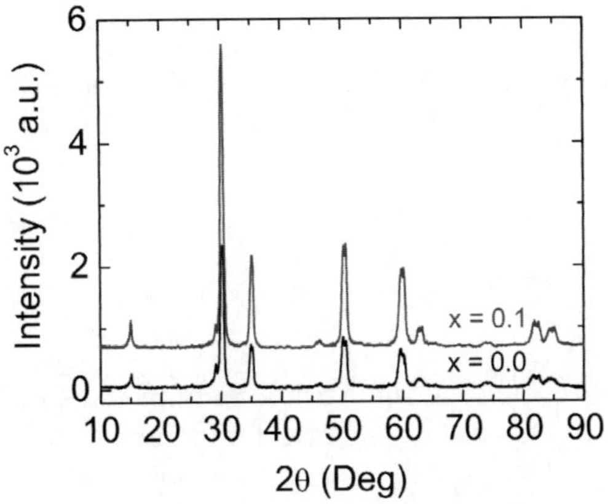


**Figure 1.** X-ray diffraction patterns of polycrystalline samples with compositions $Y_2Ir_{2x}Ru_xO_7$ series.

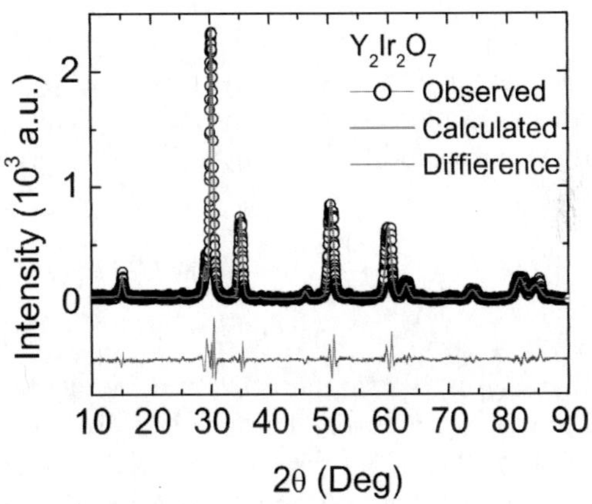


**Figure 2.** XRD pattern for $Y_2Ir_2O_7$ (x = 0) along with reitveld refinement.

## 3. Results and Discussion

### *3.1. structral properties*

Fig. 1 shows XRD pattern of $Y_2Ir_{2x}Ru_xO_7$ with x = 0.0 and 0.1 at room temperature. The figure shows there is no major modification in terms of peak splitting and/or peak positions in XRD data with Ru doping. This imply that Ru doping does not induce structural phase transition. Fig. 2 shows XRD pattern of $Y_2Ir_2O_7$ along with Rietveld refinement. The Rietveld analysis of XRD data shows the sample crystallizes in cubic structure with Fd-3m space group. The material has been found in single phase; however, a small fraction of unreacted ingredient materials is observed which is very common in this material [4, 5, 6, 9]. Considering the diamagnetic nature of ingredients, the magnetic properties of $Y_2Ir_2O_7$ are minimally influenced. The lattice constant is found to be 10.2445 and 10.2397 $\AA$ for x = 0.0 and 0.1 respectively.

### *3.2. Magnetic Properties*

Fig. 3 shows temperature dependent magnetic susceptibility of $Y_2Ir_{2x}Ru_xO_7$ measured in 10 kOe magnetic field (H) following zero field cooling (ZFC) and field cooling (FC) protocol. A bifurcation between ZFC and FC branches of magnetic susceptibility has been observed around $T_{irr}$ = 160 K and 153 K for x = 0.0 and 0.1 respectively, which implies spins are ordered below this temperature. The bifurcation between $M_{ZFC}$ and $M_{FC}$ braches could arise due to

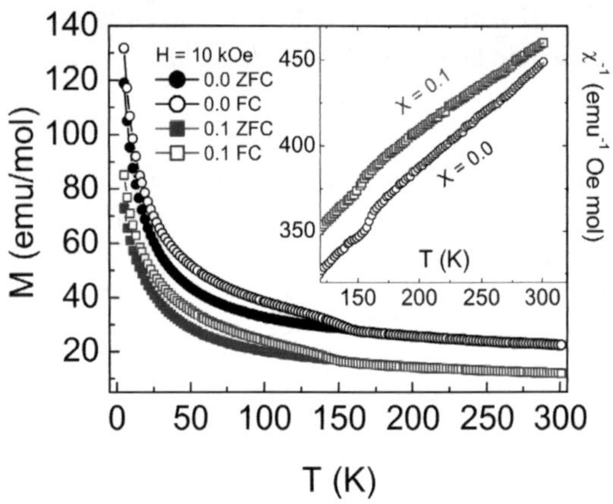


**Figure 3.** Temperature dependent dc magnetization from 5 K to 300 K at 10 kOe field (Magnetization data for x = 0.1 shifted downward by 10 (emu/mol) for clear understanding the data). Inset shows the modified CWL fitting of parent and doped sample. (discussed in text).

glassy behavior which is quite expected in pyrochlore system with geometrical frustration. The inset of Fig. 3 shows temperature dependence of inverse susceptibility ($\chi$ =M/H) which can be explained with modified Curie-Weiss law (CWL).

$$\chi = M/H = \chi_0 + C/(T - \theta_P) \tag{1}$$

Where $\chi_0$ is the temperature independent susceptibility, C is the Curie constant and $\theta_P$ is the Curie temperature. The straight line fitting of $\chi^{-1}$ data gives Curie temperature $\theta_P$ = -386 K and -335 K for x = 0.0 and 0.1 respectively, which is indicative of non-ferromagnetic (FM) behavior. The relevant frustration parameters (f) which is defined as $|\theta_p|/T_{irr}$, has been calculated to be 2.4 and 2.1 for x = 0.0 and 0.1 respectively, which suggest that frustration level is decreased with Ru-substitution.

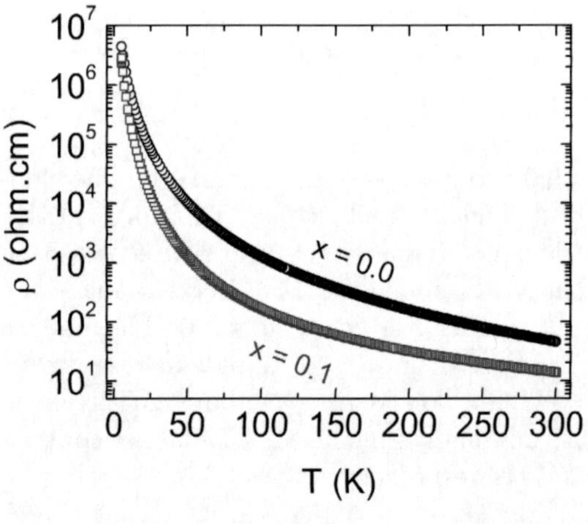


**Figure 4.** The temperature dependence electrical resistivity from 5 K to 300 K.

### *3.3. Transport properties*

Fig. 4 shows the temperature dependent electrical resistivity for $Y_2Ir_{2x}Ru_xO_7$ (x = 0.0 and 0.1). The $\rho$ (T) shows an insulating behavior for all composition which is intrinsic to most of the iridium based oxides due opening of an insulating gap [10] i.e., the resistivity increases with decreasing temperature. At low temperature below 40 K, a sudden rise in resistivity of $Y_2Ir_2O_7$ is observed where it increases by about five orders. We observed that resistivity decreases with Ru doping. However, decrease in resistivity in $Y_2Ir_{2x}Ru_xO_7$ can be attributed to tuning of SOC, $U$ with $Ru^{4+}$ ($4d^4$) substitution. Moreover, Ru electrons are itinerant in character can result in increase conduction. However, detail investigation will be done in future with more doping concentration to understand this.

## 4. Acknowledgments

We acknowledge UGC-DAE CSR, Indore, and Dr. A. Banerjee and Dr. R. Rawat for magnetization and resistivity data. We also thank to Mr. Kranti Kumar and Mr. Sachin Kumar for help in measurements. HK acknowledges UGC, India for BSR fellowship.

## References

[1] Pesin D and Balents L 2010 *Nature Phys* **6** 376
[2] Matsuhira K, Wakeshima M, Hinatsu Y and Takagi S 2011 *J. Phys. Soc. Jpn.* **80** 094701
[3] Chen G and Hermele M 2012 *Phys. Rev.* B **86** 235129
[4] Kumar H and Pramanik A K 2016 *J. Magn. Magn. Mater.* **409** 20; 2016 *AIP Conference Proceedings* **1731** 130053
[5] Disseler S M, Dhital C, Amato A, Giblin S R, de la Cruz C , Wilson S D, and Graf M J 2012 *Phys. Rev.* B **86** 014428
[6] Shapiro M C, Riggs S C, Stone M B, de la Cruz C R , Chi S, Podlesnyak A A and Fisher I R 2012 *Phys. Rev.* B **85** 214434
[7] Fukazawa H and Maeno Y 2002 *J. Phys. Soc. Jpn.* **71** 2578
[8] Taira N, Wakeshima M and Hinatsu Y 2001 *J. Phys.: Condens. Matter* **13** 5527
[9] Zhu W K, Wang M, Seradjeh B, Yang F and Zhang S X 2014 *Phys. Rev.* B **90** 054419
[10] Bhatti I N, Rawat R, Banerjee A and Pramanik A K 2014 *J. Phys.:Condens. Matter* **27** 016005; *arXiv:1512.02041 [cond-mat.str-el]*; *AIP Conference Proceedings* **1731** 110046

# Emergence of magnetic long-range order in kagome quantum antiferromagnets

**Johannes Richter and Oliver Götze**

Institut für Theoretische Physik, Universität Magdeburg, 39016 Magdeburg, Germany

E-mail: `johannes.richter@physik.uni-magdeburg.de`

**Abstract.** The existence of a spin-liquid ground state of the $s = 1/2$ Heisenberg kagome antiferromagnet (KAFM) is well established. Meanwhile, also for the $s = 1$ Heisenberg KAFM evidence for the absence of magnetic long-range order (LRO) was found. Magnetic LRO in Heisenberg KAFMs can emerge by increasing the spin quantum number $s$ to $s > 1$ and for $s = 1$ by an easy-plane anisotropy. In the present paper we discuss the route to magnetic order in $s = 1/2$ KAFMs by including an isotropic interlayer coupling (ILC) $J_\perp$ as well as an easy-plane anisotropy in the kagome layers by using the coupled-cluster method to high orders of approximation. We consider ferro- as well as antiferromagnetic $J_\perp$. To discuss the general question for the crossover from a purely two-dimensional (2D) to a quasi-2D and finally to a three-dimensional system we consider the simplest model of stacked (unshifted) kagome layers. Although the ILC of real kagome compounds is often more sophisticated, such a geometry of the ILC can be relevant for barlowite. We find that the spin-liquid ground state present for the strictly 2D $s = 1/2$ $XXZ$ KAFM survives a finite ILC, where the spin-liquid region shrinks monotonously with increasing anisotropy. If the ILC becomes large enough (about 15% of intralayer coupling for the isotropic Heisenberg case and about 4% for the $XY$ limit) magnetic LRO can be established, where the $q = 0$ symmetry is favorable if $J_\perp$ is of moderate strength. If the strength of the ILC further increases, $\sqrt{3} \times \sqrt{3}$ LRO can become favorable against $q = 0$ LRO.

## 1. Introduction

One of the most fascinating problems in frustrated quantum magnetism is the investigation of the ground state (GS) of the $s = 1/2$ antiferromagnet on the kagome lattice, see, e.g., Refs. [1–20]. The existence of a spin-liquid ground state of the $s = 1/2$ Heisenberg kagome antiferromagnet (KAFM) is well established, although the discussion of the nature of the spin-liquid state is still not settled. Meanwhile, also for the $s = 1$ Heisenberg KAFM evidence for the absence of magnetic long-range order (LRO) was found [12, 21–24]. On the other hand, recently it has been reported that magnetic $\sqrt{3}\times\sqrt{3}$ LRO can be established by increasing the spin quantum number to $s > 1$ [12, 20, 25]. Further increasing $s$ stabilizes this kind of order, thus a coupled-cluster study reports for the magnetic order parameter (sublattice magnetization) $M/s = 0.074$, 0.203, 0.294, and 0.358 for $s = 3/2$, 2, 5/2 and 3, respectively [12]. Note that in [20] for $s = 3/2$ a similar value for $M$ was reported, whereas in [25] a siginifcantly larger $M$ was obtained.

Another way to decrease quantum fluctuations is to include exchange anisotropy [18, 25–34] or to take into account further-neighbor couplings [31, 35–48]. Such modifications of the pure nearest-neighbor (NN) Heisenberg KAFM may play a crucial role to establish GS magnetic LRO of $\sqrt{3} \times \sqrt{3}$ or of $q = 0$ symmetry. For the $XXZ$ $s = 1/2$ KAFM with NN couplings only, an easy-axis anisotropy is not sufficient to establish magnetic LRO [18, 29–31, 34]. On the other hand, for $s = 1$ there is evidence [20, 28, 30] that a finite region of disorder around the isotropic Heisenberg point gives way for $\sqrt{3} \times \sqrt{3}$ LRO as increasing the easy-axis anisotropy. Further increasing the anisotropy towards the $XY$ model yields a second transition from $\sqrt{3} \times \sqrt{3}$ to $q = 0$ LRO.

An obvious route towards magnetic LRO in the GS and also at finite temperatures is given by including a coupling between kagome layers. Here we investigate the effect of an isotropic interlayer coupling (ILC) on the ground state order in a stacked $s = 1/2$ $XXZ$ KAFM with easy-plane anisotropy in the layers by using the coupled-cluster method (CCM) to high orders of approximation. The corresponding Hamiltonian reads

$$H = \sum_n \left\{ \sum_{\langle ij \rangle} \left( s^x_{i,n} s^x_{j,n} + s^y_{i,n} s^y_{j,n} + \Delta s^z_{i,n} s^z_{j,n} \right) \right\} + J_\perp \sum_{i,n} \mathbf{s}_{i,n} \cdot \mathbf{s}_{i,n+1}, \tag{1}$$

where $n$ labels the kagome layers and $J_\perp$ is a perpendicular (i.e., non-frustrated) ILC. The expression in curly brackets represents the $XXZ$ KAFM of the layer $n$ with NN intralayer couplings $J = 1$ and an easy-axis anisotropy $0 \leq \Delta \leq 1$. We do not restrict the sign of the ILC, i.e., we consider ferro- as well as antiferromagnetic $J_\perp$.

It is in order to notice that the ILC in real kagome compounds is often more sophisticated than that one considered here. Hence, the aim of our paper is not the discussion of particular kagome compounds, rather it is a theoretical investigation of the general question for the crossover from a purely two-dimensional (2D) to a quasi-2D and then to a three-dimensional system. For that it is appropriate to consider the simplest model of stacked unshifted kagome layers with a perpendicular ILC. There is at least one kagome compound, namely barlowite, with unshifted kagome layers, see, e.g. [49, 50]. As it has been pointed out very recently [51], through isoelectronic substitution in barlowite this kagome system becomes quite similar to our model system.

As already mentioned above, for isolated $s = 1/2$ kagome layers the GS is always magnetically disordered. Very recently, for the fully isotropic model it has been found that this non-magnetic ground state persists until relatively large strengths of the ILC [52], i.e., the critical strength of the ILC $J_\perp^c$ needed to establish magnetic LRO is about 15% of the intralayer coupling. The question arises how an additional $XXZ$ anisotropy in the layers influences $J_\perp^c$. Does the magnetically disordered GS survive also for the extreme limit of $XY$ layers, if the ILC becomes nonzero? Bearing in mind that the GS selection for the strictly 2D model depends on the anisotropy parameter $\Delta$ [20, 28, 30] we may also ask which GS symmetry ($\sqrt{3} \times \sqrt{3}$ or of $q = 0$) is selected when including the ILC. Moreover, we will discuss to what extent the sign of $J_\perp$ is relevant.

## 2. Brief outline of the coupled-cluster method (CCM)

The CCM is meanwhile a well-established and well-tested method in frustrated quantum magnetism, see, e.g. [12, 30, 37, 39, 52–63]. Hence, we want to illustrate here only some basic features of the CCM. At that we follow [12, 30], where the CCM was applied to the 2D KAFM, and [52], where the isotropic stacked $s = 1/2$ KAFM was studied by means of the CCM. For more general information on the CCM, see, [64–69]. Note first that the CCM yields results directly for number of sites $N \rightarrow \infty$. The starting point of the CCM calculation is a normalized reference state $|\Phi\rangle$. From a quasi-classical perspective that is the stacked coplanar $\sqrt{3} \times \sqrt{3}$ state or $q = 0$ state (see, e.g., Refs. [28, 30, 70–73]). Then we perform a rotation of the local axes of each of the spins such that all spins in the reference state align along the negative $z$ axis. Within this rotated local spin coordinate system we define a complete set of multispin creation operators $C_I^+ \equiv (C_I^-)^\dagger$ related to this reference state:

$|\Phi\rangle = |\downarrow\downarrow\downarrow\cdots\rangle$; $C_I^+ = s_n^+\,,\, s_n^+ s_m^+\,,\, s_n^+ s_m^+ s_k^+\,,\,\ldots$ ,, i.e., the spin operators entering the multispin creation operators are defined in the local coordinate frames. The indices $n, m, k, \ldots$ denote arbitrary lattice sites. The ket and bra GS eigenvectors $|\Psi\rangle$ and $\langle\tilde{\Psi}|$ of the spin system are given by $|\Psi\rangle = e^S|\Phi\rangle$ , $S = \sum_{I\neq 0} a_I C_I^+$ ; $\langle\tilde{\Psi}| = \langle\Phi|\tilde{S}e^{-S}$ , $\tilde{S} = 1 + \sum_{I\neq 0}\tilde{a}_I C_I^-$. For the coefficients $a_I$ and $\tilde{a}_I$ in the CCM correlation operators, $S$ and $\tilde{S}$, the ket-state and bra-state equations $\langle\Phi|C_I^- e^{-S} H e^S|\Phi\rangle = 0$ , $\langle\Phi|\tilde{S}e^{-S}[H, C_I^+]e^S|\Phi\rangle = 0$ , $\forall I \neq 0$, hold, where each equation belongs to a certain multispin configuration $I$, i.e., to a certain configuration of lattice sites $n, m, k, \ldots$ . For the GS energy $E_0 = \langle\Phi|e^{-S}He^S|\Phi\rangle$ holds. The sublattice magnetization (magnetic order parameter) is given by $M = -\frac{1}{N}\sum_{i=1}^{N}\langle\tilde{\Psi}|s_i^z|\Psi\rangle$, where $s_i^z$ is expressed in the rotated local coordinate system. For the many-body problem at hand, we have to truncate the expansions of $S$ and $\tilde{S}$. For that we use the standard LSUB$m$ approximation scheme, cf., e.g., [12, 30, 37, 39, 52–58, 60–63, 67, 68], where no more than $m$ spin flips spanning a range of no more than $m$ contiguous lattice sites are included. Using the parallelized CCM code [74] we can solve the set of CCM equations up to LSUB8. For this approximation level the number of CCM equations is 278,024. We then extrapolate the 'raw' LSUB$m$ data to the limit $m \to \infty$. In correspondence to [52] we use two schemes, namely an extrapolation using $m = 4, 5, \ldots, 8$ (scheme I) and separately an extrapolation using $m = 4, 6, 8$ (scheme II). Scheme I corresponds to that one used for the 2D KAFM [12, 30], whereas scheme II (not including the odd LSUB$m$ approximation levels) is more appropriate for magnets with collinear AFM correlations [53–55, 57–59, 62, 65, 75]. By comparing the results of schemes I and II we can get information on the precision of the extrapolated data.

For the GS energy the formula $e_0(m) = E_0(m)/N = e_0(m \to \infty) + a_1/m^2 + a_2/m^4$ is well tested and it provides precise data for the extrapolated energy $e_0(m \to \infty)$; for the magnetic order parameter $M$ the ansatz $M(m) = M(m \to \infty) + b_1(1/m)^{1/2} + b_2(1/m)^{3/2}$ is an appropriate choice to determine quantum critical points for frustrated quantum spin systems [12, 30, 37, 39, 52–55, 57, 58, 62, 63].

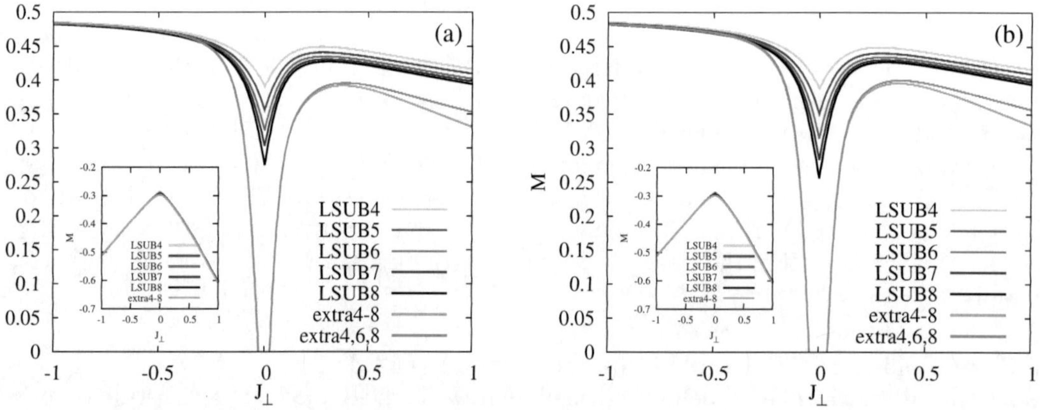


**Figure 1.** CCM-LSUB$m$ as well as extrapolated GS sublattice magnetization $M$ for $\Delta = 0$ ($XY$ limit) using (a) the $\sqrt{3}\times\sqrt{3}$ reference state and (b) the $q = 0$ reference state as a function of the ILC $J_\perp$. The labels 'extra4-8' and 'extra4,6,8' correspond to the extrapolation schemes I and II, respectively (see main text). Inset: Corresponding data for the GS energy $e_0$. Note that for $e_0$ the results of both extrapolation schemes are practically identical, and, therefore we show only extrapolated data for scheme I.

## 3. Results

We start with a discussion of the most anisotropic $XY$ limit, i.e., $\Delta = 0$. For that case we present in Figs. 1(a) and (b) the full set of LSUB$m$ data as well as the extrapolated data for the order parameter $M$ (main panels) and also for the GS energy per spin $e_0 = E_0/N$ (insets) for the $\sqrt{3} \times \sqrt{3}$ and $q = 0$ reference states, respectively. (Note that corresponding figures for $\Delta = 1$ can be found in [52].) It is evident that $e_0$ converges quickly as the level $m$ of the LSUB$m$ approximation increases. As a result, the extrapolation with leading order $1/m^2$ is very accurate, as it has been demonstrated in many cases, where data from other precise methods are available to compare with, see, e.g., Refs. [12,30,65]. Obviously, the shape of the curves and the magnitude of the energies are very similar for both states, where from [30] it is known that for the 2D $XY$ model, i.e., at $J_\perp = 0$ and $\Delta = 0$, the $q = 0$ state has slightly lower energy. As found previously [18, 29–31, 34], $M$ is zero for $J_\perp = 0$. For $\Delta = 0$ the critical ILCs, where magnetic LRO sets in, are : $J_\perp^c = -0.027$, $J_\perp^c = +0.026$ ($\sqrt{3} \times \sqrt{3}$ state) and $J_\perp^c = -0.044$, $J_\perp^c = +0.042$ ($q = 0$ state) when using scheme I, and, $J_\perp^c = -0.028$, $J_\perp^c = +0.028$ ($\sqrt{3} \times \sqrt{3}$ state) and $J_\perp^c = -0.044$, $J_\perp^c = +0.042$ ($q = 0$ state) when using scheme II. Corresponding values for $\Delta = 1$ are [52]: $J_\perp^c = -0.100$, $J_\perp^c = +0.102$ ($\sqrt{3} \times \sqrt{3}$ state) and $J_\perp^c = -0.154$, $J_\perp^c = +0.151$ ($q = 0$ state) for scheme I, and, $J_\perp^c = -0.104$, $J_\perp^c = +0.110$ ($\sqrt{3} \times \sqrt{3}$ state) and $J_\perp^c = -0.135$, $J_\perp^c = +0.130$ ($q = 0$ state) for scheme II. We may conclude that (i) there is a good agreement of the critical ILCs obtained by both extrapolation schemes, (ii) the spin-liquid region survives a finite ILC even in the $XY$ limit, and (iii) due to anisotropy this region becomes noticeably smaller.

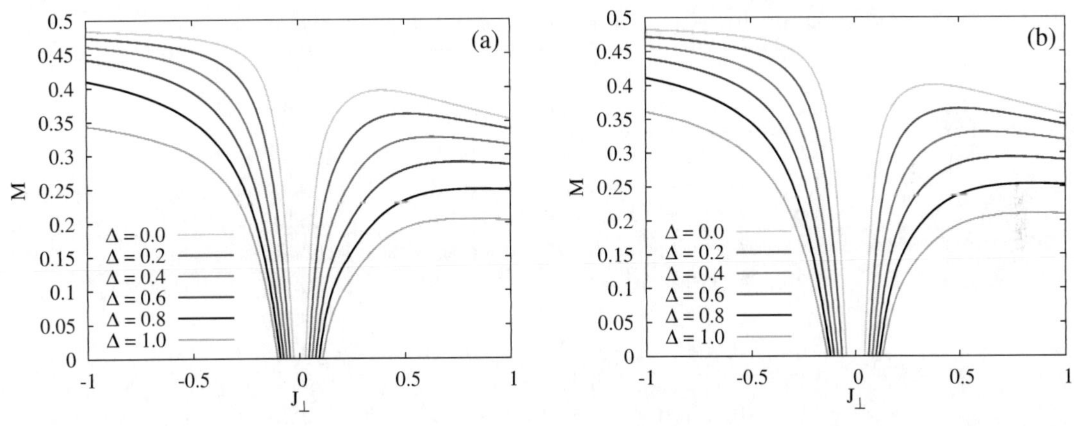


**Figure 2.** Extrapolated GS sublattice magnetization $M$ (scheme II) for various values of the anisotropy parameter $\Delta =$ using (a) the $\sqrt{3} \times \sqrt{3}$ reference state and (b) the $q = 0$ reference state as a function of the ILC $J_\perp$.

In the next figure we compare the extrapolated magnetic order parameter $M$ for several values of the anisotropy parameter $\Delta$, see Figs. 2 (a) and (b). Because (at least for small values of the $|J_\perp|$) both extrapolation schemes lead to very similar results, for convenience, we show only the data of scheme II. The critical values of the ILC, where magnetic LRO sets in, are listed for both extrapolation schemes in Tables 1 and 2. From those tables and from Figs. 2 (a) and (b) we conclude: (i) for all values $0 \leq \Delta \leq 1$ there is a finite region of magnetic disorder, (ii) this region shrinks monotonously with increasing anisotropy (i.e. with decreasing $\Delta$), (iii) near the critical ILC $J_\perp^c$ the downturn of the magnetic order parameter $M$ becomes very steep as increasing the anisotropy, (iv) in the magnetically ordered phase the sublattice magnetization

**Table 1.** Critical values $J_\perp^c$ of the ILC where magnetic LRO emerges for antiferromagnetic $J_\perp$ in dependence on the anisotropy parameter $\Delta$. The labels 'extra4-8' and 'extra4,6,8' correspond to the extrapolation schemes I and II, respectively (see main text).

| | $q=0$ | | $\sqrt{3}\times\sqrt{3}$ | |
|---|---|---|---|---|
| $\Delta$ | extra4,6,8 | extra4-8 | extra4,6,8 | extra4-8 |
| 0.0 | 0.042 | 0.042 | 0.028 | 0.026 |
| 0.1 | 0.052 | 0.052 | 0.035 | 0.032 |
| 0.2 | 0.062 | 0.062 | 0.043 | 0.039 |
| 0.3 | 0.071 | 0.073 | 0.052 | 0.046 |
| 0.4 | 0.080 | 0.083 | 0.059 | 0.053 |
| 0.5 | 0.089 | 0.094 | 0.068 | 0.060 |
| 0.6 | 0.097 | 0.104 | 0.076 | 0.068 |
| 0.7 | 0.106 | 0.115 | 0.084 | 0.076 |
| 0.8 | 0.114 | 0.127 | 0.093 | 0.085 |
| 0.9 | 0.122 | 0.138 | 0.101 | 0.094 |
| 1.0 | 0.130 | 0.151 | 0.110 | 0.102 |

**Table 2.** Critical values $J_\perp^c$ of the ILC where magnetic LRO emerges for ferromagnetic $J_\perp$ in dependence on the anisotropy parameter $\Delta$. The labels 'extra4-8' and 'extra4,6,8' correspond to the extrapolation schemes I and II, respectively (see main text).

| | $q=0$ | | $\sqrt{3}\times\sqrt{3}$ | |
|---|---|---|---|---|
| $\Delta$ | extra4,6,8 | extra4-8 | extra4,6,8 | extra4-8 |
| 0.0 | −0.044 | −0.044 | −0.029 | −0.027 |
| 0.1 | −0.055 | −0.055 | −0.037 | −0.034 |
| 0.2 | −0.065 | −0.066 | −0.045 | −0.040 |
| 0.3 | −0.075 | −0.077 | −0.053 | −0.047 |
| 0.4 | −0.085 | −0.088 | −0.061 | −0.054 |
| 0.5 | −0.094 | −0.099 | −0.069 | −0.062 |
| 0.6 | −0.104 | −0.110 | −0.076 | −0.070 |
| 0.7 | −0.112 | −0.121 | −0.083 | −0.077 |
| 0.8 | −0.120 | −0.132 | −0.091 | −0.085 |
| 0.9 | −0.128 | −0.143 | −0.098 | −0.093 |
| 1.0 | −0.135 | −0.154 | −0.104 | −0.100 |

$M$ grows noticeably as increasing the anisotropy, and, (v) while for ferromagnetic $J_\perp < 0$ there is a monotonic increase of $M$ with increasing $|J_\perp|$, for $J_\perp > 0$ we find a non-monotonic behavior of $M$. Interestingly, for smaller values of $\Delta$ the maximum in $M$ appears at values of the ILC significantly before $J_\perp = 1$. For $\Delta = 0$ it is at about $J_\perp = 0.36$ for both reference states.

Now we address the question which magnetic GS LRO is selected by quantum fluctuations. As we know from previous studies, the mechanism of quantum selection of the GS LRO in the KAFM is very subtle, and the energy difference between competing states can be very small [28, 30, 52]. The CCM approach in high orders of approximation provides an accurate tool to compare these energies. Thus, the CCM for $s > 1/2$ yields the correct quantum selection of the $\sqrt{3} \times \sqrt{3}$ GS vs. the $q = 0$ GS as obtained by non-linear spin-wave theory [12]. A direct comparison of CCM and non-linear spin-wave data for energy differences (which are also of the order $10^{-3} \ldots 10^{-4}$) for the $XXZ$ KAFM for large $s$, given in Fig. 3 of Ref. [30], demonstrates that both independent approaches agree very well. In Fig. 3 we show our results for the energy difference $\delta e = e_0^{\sqrt{3}\times\sqrt{3}} - e_0^{q=0}$ as a function of $J_\perp$ for various values of $\Delta$. We mention first that both extrapolation schemes I and II yield consistent results for $\delta e$. Therefore, for convenience, we show in Fig. 3 only the data of scheme II. From Fig. 3 it is evident that at low values of $|J_\perp|$ the $q = 0$ reference state yields lower energy, i.e. $\delta e > 0$. That is consistent with Refs. [12] and [30], where the case $J_\perp = 0$ was studied. For ferro- as well as antiferromagnetic ILC $\delta e$ is still positive at those values of $J_\perp$, where the sublattice magnetizations $M_{\sqrt{3}\times\sqrt{3}}$ or $M_{q=0}$ become larger than zero. Hence, our results demonstrate that a magnetic disorder-to-order transition to $q = 0$ LRO takes place at $J_\perp^c$ as listed in columns 2 and 3 of Tables 1 and 2. We mention that the quantum selection of the $q = 0$ GS LRO is contrary to the semi-classical large-$s$ selection of the $\sqrt{3} \times \sqrt{3}$ LRO found for the 2D spin-$s$ KAFM. As already found for the isotropic case, $\Delta = 1$, further increasing the strength of $J_\perp$ can lead to a second transition at $J_\perp^t$ from $q = 0$ to $\sqrt{3} \times \sqrt{3}$ LRO determined by the change of sign of $\delta e$. Here, we find such a second transition for all values $\Delta$ on the antiferromagnetic side, $J_\perp > 0$. On the ferromagnetic side a second transition is found only for $\Delta = 1, 0.9, 0.8$. For lower values of $\Delta$ the energy difference $\delta e$ remains positive. However, it is in order to mention that already for $\Delta < 0.9$ our CCM result for $\delta e$ becomes extremely small as $J_\perp \to -1$, and the $q = 0$ and $\sqrt{3} \times \sqrt{3}$ states are practically degenerate. We list or results for the position of the second transition between $q = 0$ and $\sqrt{3} \times \sqrt{3}$ magnetic LRO in Table 3. Summarizing our data for the transitions at $J_\perp^c$ and $J_\perp^t$ as given in Tables 1-3 we present in Fig. 4 a sketch of a phase diagram of the stacked $s = 1/2$ $XXZ$ KAFM with easy-plane anisotropy in the layers.

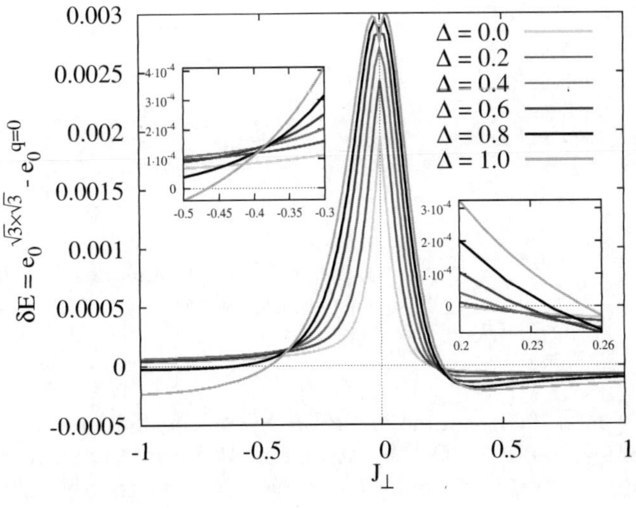


**Figure 3.** Difference $\delta e = e_0^{\sqrt{3}\times\sqrt{3}} - e_0^{q=0}$ of the extrapolated GS energies (using scheme II) of the $\sqrt{3}\times\sqrt{3}$ and the $q = 0$ states as a function of the ILC $J_\perp$. In the insets we show $\delta e$ in an enlarged scale for those regions of $J_\perp$, where $\delta e$ changes its sign.

## 4. Concluding remarks

A main finding of our paper is that the spin-liquid GS present for the strictly 2D $s = 1/2$ $XXZ$ KAFM survives a finite ILC even in the $XY$ limit of maximal easy-plane anisotropy. If the ILC becomes large enough (about 15% of the intralayer coupling in our model system for the isotropic Heisenberg case and about 4.5% for coupled $XY$ kagome layers) magnetic LRO can be established, where the $q = 0$ symmetry is favorable, if $J_\perp$ is of moderate strength. The quantum

**Table 3.** Transition points $J_\perp^t$ between $q = 0$ and $\sqrt{3} \times \sqrt{3}$ magnetic LRO. For $\Delta < 0.75$ and ferromagnetic $J_\perp$ we do find a transition.

| | AFM | | FM | |
|---|---|---|---|---|
| $\Delta$ | extra4,6,8 | extra4-8 | extra4,6,8 | extra4-8 |
| 0.0 | 0.199 | 0.220 | — | — |
| 0.1 | 0.205 | 0.234 | — | — |
| 0.2 | 0.209 | 0.242 | — | — |
| 0.3 | 0.212 | 0.248 | — | — |
| 0.4 | 0.216 | 0.254 | — | — |
| 0.5 | 0.220 | 0.259 | — | — |
| 0.6 | 0.226 | 0.265 | — | — |
| 0.7 | 0.232 | 0.273 | — | — |
| 0.8 | 0.238 | 0.282 | −0.614 | −0.562 |
| 0.9 | 0.245 | 0.294 | −0.502 | −0.466 |
| 1.0 | 0.252 | 0.310 | −0.467 | −0.435 |

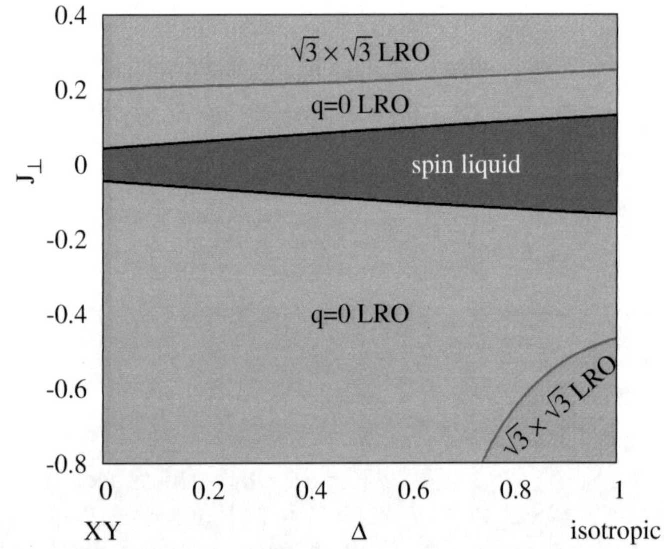


**Figure 4.** Sketch of a phase diagram of the stacked $s = 1/2$ kagome antiferromagnet with easy-plane anisotropy in the layers.

selection of the $\sqrt{3} \times \sqrt{3}$ magnetic LRO versus the $q = 0$ magnetic LRO found for larger values of antiferromagnetic $J_\perp$ and also for ferromagnetic $J_\perp$ near the isotropic limit is related to a very small energy gain. Thus, in real compounds even very small additional terms in the relevant spin Hamiltonian such as further distance exchange couplings may therefore be more relevant.

Let us finally discuss the relevance of our results for experiments on kagome compounds. As it has been pointed out very recently, through isoelectronic substitution in barlowite this kagome compound is related to our model system of stacked (unshifted) kagome layers [51]. The finding that the spin-liquid phase survives a finite ILC is relevant for the modified barlowite system, where an ILC of about 6-7% of the intralayer coupling was predicted [51]. The relation to herbertsmithite, for which an ILC of about 5% was determined [76], is only on a qualitative level, since here the kagome layers are shifted from layer to layer. The observation of $q = 0$ magnetic order in $Cs_2Cu_3SnF_{12}$ reported in [77, 78] is also in qualitative agreement with our findings for the GS selection. We also mention that anisotropic kagome antiferromagnets have relevance for the experimental research, see, e.g., Refs. [77, 79–81]. Moreover, anisotropic spin models are of great interest with respect to engineering models of quantum magnetism on optical lattices, see, e.g., [82–84].

## References


[1] R. R. P. Singh and D. A. Huse, *Phys. Rev. Lett.* **68**, 1766 (1992).
[2] C. Waldtmann et al., *Eur. Phys. J. B* **2**, 501 (1998).
[3] S. Capponi, A. Läuchli, and M. Mambrini, *Phys. Rev. B* **70**, 104424 (2004).
[4] R. R. P. Singh and D. A. Huse, *Phys. Rev. B* **76**, 180407(R) (2007).
[5] P. Sindzingre and C. Lhuillier, *Europhys. Lett.* **88**, 27009 (2009).
[6] G. Evenbly and G. Vidal, *Phys. Rev. Lett.* **104**, 187203 (2010).
[7] S. Yan, D. A. Huse, and S. R. White, *Science* **332**, 1173 (2011).
[8] H. Nakano and T. Sakai, *J. Phys. Soc. Jpn.* **80**, 053704 (2011).
[9] A. M. Läuchli, J. Sudan, and E. S. Sørensen, *Phys. Rev. B* **83**, 212401 (2011).
[10] Y.-M. Lu, Y. Ran, and P. A. Lee, *Phys. Rev. B* **83**, 224413 (2011).
[11] Y. Iqbal, F. Becca, and D. Poilblanc, *Phys. Rev. B* **84**, 020407(R) (2011).
[12] O. Götze et al., *Phys. Rev. B* **84**, 224428 (2011).
[13] S. Depenbrock, I. P. McCulloch, and U. Schollwöck, *Phys. Rev. Lett.* **109**, 067201 (2012).
[14] Y. Iqbal, F. Becca, S. Sorella, D. Poilblanc, *Phys. Rev. B* **87**, 060405(R) (2013).
[15] I. Rousochatzakis, R. Moessner, J. van den Brink, *Phys. Rev. B* **88**, 195109 (2013).
[16] Z. Y. Xie et al., *Phys. Rev. X* **4**, 011025 (2014).
[17] I. Rousochatzakis, Y. Wan, O. Tchernyshyov, and F. Mila, *Phys. Rev. B* **90**, 100406(R) (2014).
[18] W.-J. Hu, S.S. Gong, F. Becca, and D.N. Sheng, *Phys. Rev. B* **92**, 2015(R) (2015).
[19] M. P. Zaletel and A. Vishwanath, *Phys. Rev. Lett.* **114**, 077201 (2015).
[20] J. Oitmaa and R. R. P. Singh, *Phys. Rev. B* **93**, 014424 (2016)
[21] T. Liu, W. Li, A. Weichselbaum, J. von Delft, and G. Su, *Phys. Rev. B* 91, **060403** (2015).
[22] H. J. Changlani and A. M. Läuchli, *Phys. Rev. B* **91**, 100407 (2015).
[23] S. Nishimoto and M. Nakamura, *Phys. Rev. B* **92**, 140412(R) (2015).
[24] W. Li, A. Weichselbaum, J. von Delft, and H.-H. Tu, *Phys. Rev. B* **91**, 224414 (2015).
[25] Tao Liu, Wei Li, Gang Su, arXiv:1603.01935v1.
[26] O. Cepas, C. M. Fong, P. W. Leung, C. Lhuillier, *Phys. Rev. B* **78**, 140405(R) (2008).
[27] I. Rousochatzakis, S. R. Manmana, A. M. Läuchli, B. Normand, and F. Mila, *Phys. Rev. B* **79**, 214415 (2009).
[28] A. L. Chernyshev and M. E. Zhitomirsky, *Phys. Rev. Lett.* **113**, 237202 (2014).
[29] Y.C. He and Y. Chen, *Phys. Rev. Lett.* **114**, 037201 (2015)
[30] O. Götze and J. Richter, *Phys. Rev. B* **91**, 104402 (2015).
[31] W. Zhu, S. S. Gong, and D. N. Sheng, *Phys. Rev. B* **92**, 014424 (2015).
[32] K. Kumar, K. Sun, and E. Fradkin, *Phys. Rev. B* **92**, 094433 (2015).
[33] A. L. Chernyshev and M. E. Zhitomirsky, *Phys. Rev. B* **92**, 144415 (2015).
[34] K. Essafi, O. Benton, and L.D.C. Jaubert, *Nat. Commun.* **7**, 10297 (2016)
[35] J.-C. Domenge, P. Sindzingre, C. Lhuillier, and L. Pierre, *Phys. Rev. B* **72**, 024433 (2005).
[36] O. Janson, J. Richter, and H. Rosner, *Phys. Rev. Lett.* **101**, 106403 (2008).
[37] R.F. Bishop, P.H.Y. Li, D.J.J. Farnell, and C.E. Campbell, *Phys. Rev. B* **82**, 104406 (2010).
[38] T. Tay and O. I. Motrunich, *Phys. Rev. B* **84**, 020404(R) (2011).
[39] P.H.Y. Li, R.F. Bishop, C.E. Campbell, D.J.J. Farnell, O. Götze, and J. Richter, *Phys. Rev. B* **86**, 214403 (2012).
[40] H.-C. Jiang, Z. Wang, and L. Balents, *Nature Physics* **8**, 902 (2012).
[41] B. Bauer et al., *Nature Communications* **5**, 5137 (2014).
[42] R. Suttner, C. Platt, J. Reuther, and R. Thomale, *Phys. Rev. B* **89**, 020408(R) (2014).
[43] W.J. Hu, W. Zhu, Y. Zhang, S.S. Gong, F. Becca, and D.N. Sheng *Phys. Rev. B* **91**, 041124(R) (2015)
[44] S.S. Gong, W. Zhu, L. Balents, and D. N. Sheng *Phys. Rev. B* **91**, 075112 (2015).
[45] F. Kolley, S. Depenbrock, I. P. McCulloch, U. Schollwöck, and V. Alba, *Phys. Rev. B* **91**, 104418 (2015).
[46] S. Bieri, L. Messio, B. Bernu, and C. Lhuillier, *Phys. Rev. B* **92**, 060407(R) (2015).
[47] A. Wietek, A.Sterdyniak, and A. M. Laeuchli, *Phys. Rev. B* **92**, 125122 (2015).
[48] Y. Iqbal et al., *Phys. Rev. B* **92**, 220404 (2015).
[49] T.-H. Han, J. Singleton, J. A. Schlueter, *Phys. Rev. Lett.* **113**, 227203 (2014).
[50] H. O. Jeschke et al., *Phys. Rev. B* **92**, 094417 (2015).

[51] D. Guterding, R. Valenti, and H. O. Jeschke, arXiv:1605.08162.
[52] O. Götze and J. Richter, *Europhys. Lett. (EPL)* **114**, 67004 (2016).
[53] D. Schmalfuß, R. Darradi, J. Richter, J. Schulenburg, and D. Ihle, *Phys. Rev. Lett.* **97**, 157201 (2006).
[54] R. Darradi, O. Derzhko, R. Zinke, J. Schulenburg, S.E. Krüger and J. Richter, *Phys. Rev. B* **78**, 214415 (2008).
[55] R. Zinke, J. Schulenburg, and J. Richter, *Eur. Phys. J. B* **61**, 147 (2008).
[56] D.J.J. Farnell, R. Zinke, J. Schulenburg, and J. Richter, *J. Phys.: Cond. Matter* **21**, 406002 (2009).
[57] J. Richter, R. Darradi, J. Schulenburg, D.J.J. Farnell, and H. Rosner, *Phys. Rev. B* **81**, 174429 (2010).
[58] D.J.J. Farnell et al., *Phys. Rev. B* **84**, 012403 (2011).
[59] D.J.J. Farnell et al., *Phys. Rev. B* **89**, 184407 (2014).
[60] R. F. Bishop, P. H. Y. Li, and C. E. Campbell, *Phys. Rev. B* **89**, 214413 (2014).
[61] J.-J. Jiang, Y.-J Liu, F. Tang, C.-H. Yang, and Y.-B. Sheng, *Physica B: Cond. Mat.* **463**, 30 (2015).
[62] J. Richter, R. Zinke, D.J.J. Farnell, *Eur. Phys. J. B* **88**, 2 (2015).
[63] P. H. Y. Li, R. F. Bishop, and C. E. Campbell, *Phys. Rev. B* **91**, 014426 (2015).
[64] R.F. Bishop, in *Microscopic Quantum Many-Body Theories and Their Applications*, edited by J. Navarro and A. Polls, *Lecture Notes in Physics* **510** (Springer, Berlin, 1998), p.1.
[65] D.J.J. Farnell and R.F. Bishop, in *Quantum Magnetism, Lecture Notes in Physics* **645**, 307 (2004)
[66] R.F. Bishop, D.J.J. Farnell, S.E. Krüger, J.B. Parkinson, *J. Phys.: Condens. Matter* **12**, 6887 (2000).
[67] M. Roger and J.H. Hetherington, *Phys. Rev. B* **41**, 200 (1990).
[68] R.F. Bishop, J.B. Parkinson, and Y. Xian, *Phys. Rev. B* **43**, R13782 (1991); *Phys. Rev. B* **44**, 9425 (1991).
[69] C. Zeng, D.J.J. Farnell, and R.F. Bishop, *J. Stat. Phys.* **90**, 327 (1998).
[70] A. B. Harris, C. Kallin and A. J. Berlinsky, *Phys. Rev. B* **45**, 2899 (1992).
[71] S. Sachdev, *Phys. Rev. B* **45**, 12377 (1992).
[72] A. Chubukov, *Phys. Rev. Lett.* **69**, 832 (1992).
[73] C. L. Henley and E. P. Chan, *J. Magn. Magn. Mater.* **140-144**, 1693 (1995).
[74] For the numerical calculation we use the program package 'The crystallographic CCM' (D.J.J. Farnell and J. Schulenburg).
[75] D. J. J. Farnell and R. F. Bishop, *Int. J. Mod. Phys. B* **22**, 3369 (2008).
[76] O. Janson, *DFT based microscopic magnetic modeling for low-dimensional spin systems*, Ph.D. thesis, Technische Universität Dresden (2012)
[77] T. Ono, K. Matan, Y. Nambu, T. J. Sato, K. Katayama, S. Hirata, and H. Tanaka, J. Phys. Soc. Jpn. **83**, 043701 (2014); K. Katayama, N. Kurita, H. Tanaka, arXiv:1412.5770.
[78] K. Katayama, N. Kurita, and H. Tanaka, *Phys. Rev. B* **91**, 214429 (2015).
[79] K. Matan, D. Grohol, D. G. Nocera, T. Yildirim, A. B. Harris, S. H. Lee, S. E. Nagler, and Y. S. Lee, Phys. Rev. Lett. **96**, 247201 (2006).
[80] A. Zorko, S. Nellutla, J. van Tol, L. C. Brunel, F. Bert, F. Duc, J.-C. Trombe, M. A. de Vries, A. Harrison, and P. Mendels, Phys. Rev. Lett. **101**, 026405 (2008).
[81] A. Zorko, F. Bert, A. Ozarowski, J. van Tol, D. Boldrin, A. S. Wills, and P. Mendels, Phys. Rev. B **88**, 144419 (2013).
[82] S. V. Isakov, S. Wessel, R. G. Melko, K. Sengupta, Yong Baek Kim, Phys. Rev. Lett. **97**, 147202 (2006).
[83] I. Bloch, J. Dalibard, and W. Zwerger, Rev. Mod. Phys. **80**. 885 (2008).
[84] J. Struck, M. Weinberg, C. Ölschläger, P. Windpassinger, J. Simonet, K. Sengstock, R. Höppner, P. Hauke, A. Eckardt, M. Lewenstein, and L. Mathey, Nature Physics **9**, 738 (2013).

IOP Conf. Series: Journal of Physics: Conf. Series **828** (2017) 012011 doi:10.1088/1742-6596/828/1/012011

# High-field $^{1}$H-NMR study around a 1/4 plateau of quantum spin dimer system $NH_4CuCl_3$

**K Matsui [1,4], M Fujisawa [2], H Tanaka [2], R Scheuermann [3] and T Goto [1]**

[1]Physics Division, Sophia University, Tokyo 102-8544, Japan.
[2]Department of Physics, Tokyo Institute of Technology, Meguro-ku, Tokyo 152-8551, Japan.
[3]Paul Scherrer Institute, CH-5232 Villigen PSI, Switzerland.
[4]Research Fellow of Japan Society for the Promotion of Science.

E-mail: k703861@eagle.sophia.ac.jp

**Abstract.** We have investigated high-field TF-$\mu$SR and $^{1}$H-NMR on the $S = 1/2$ three dimensional spin dimer system $NH_4CuCl_3$, which shows a magnetic order at $T_N = 1.29$ K under zero field and the two-stepped plateaus in high field regions $H_{C1} = 5.0$ T – $H_{C2} = 12.8$ T and $H_{C3} = 17.9$ T – $H_{C4} = 24.7$ T. Both probes showed spectra containing multiple peaks corresponding to different hyperfine fields, demonstrating the existence of magnetically inequivalent dimers in a unit cell. In the slope-state field region $H < H_{C1}$, the existence of magnetic order was confirmed by the splitting in some of the NMR peaks. The whole width of NMR and $\mu$SR spectra increased with increasing field, and no drastic change in their profile was observed at $H_{C1}$. This indicates that the change in the spin state is continuous on entering the plateau state, and that no liquid-solid transition of magnons takes place at the phase boundary.

## 1. Introduction

The magnetization process of quantum spin systems is attracting much interest, because some systems show the magnetization plateaus which cannot be understood in terms of the conventional classical spin vector models [1]. $NH_4CuCl_3$ is an $S = 1/2$ three dimensional spin dimer system, which shows two-stepped magnetization plateaus at 1/4 and 3/4 of the saturation magnetization. The plateaus appear for any field directions, indicating that its mechanism roots in the quantum effect. At zero field, $NH_4CuCl_3$ does not have a spin excitation gap, and shows a magnetic order at $T_N = 1.29$ K [2, 3], which slightly increases with increasing field, takes maximum of 1.6 K at 2.7 T, and decreases to zero at $H_{C1} = 5.0$ T, as shown in Fig. 1(a). The 1/4 plateau appears in the field region between $H_{C1}$ and $H_{C2} = 12.8$ T, and the 3/4 plateau, between $H_{C3} = 17.9$ T and $H_{C4} = 24.7$ T for the field direction $H \parallel b$-axis [4]. So far, to explain the mechanism of plateaus, many theoretical models have been proposed [5, 6]. Among them, Matsumoto [7] and Rüegg [3] proposed independently the inequivalent dimer model, in which they suggest the existence of a gapless dimer and the other two dimers with different gaps. Although, $^{63/65}$Cu- and $^{35/37}$Cl-NMR [8] and $\mu$SR experiments [9] support the existence of such inequivalent dimers in the paramagnetic state below 70 K, detailed spin states in and out of the plateau-field region are still veiled. Another interesting topic in this compound is the possibility of change anomaly due to a non-collinear spin structure in the ordered state. In the magnon Bose-Einstein condensation (BEC) for the spin dimer system such as $TlCuCl_3$ [10, 11],

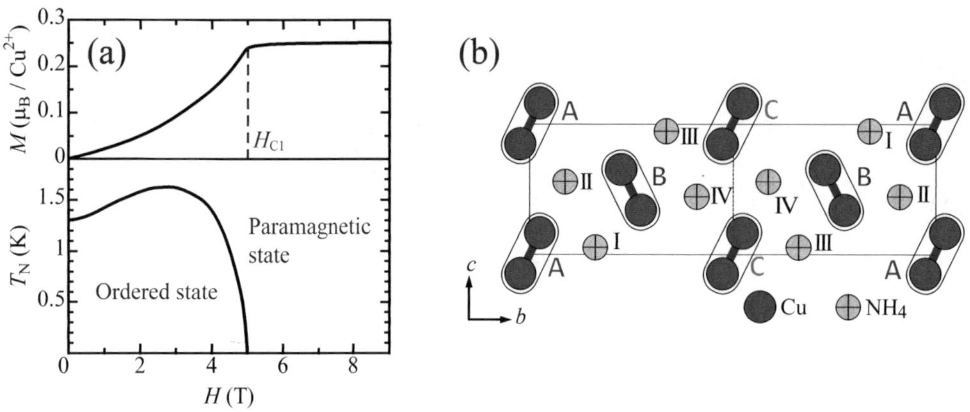


**Figure 1.** (a) Schematic phase diagram based on specific heat [2] and neutron scattering [3] (below) and magnetization curve in $H \parallel b$-axis [4] (top). (b) Schematic crystal structure of $NH_4CuCl_3$ below 70 K, where $b$-axis is doubled due to structural phase transition [3].

which is isomorphic to $NH_4CuCl_3$, the magnetic order within the moment perpendicular to the applied field is accompanied by the charge and lattice distortion [12, 13]. These can possibly be found in the present system. In order to investigate the spin state in the paramagnetic and ordered state microscopically, we have performed $^1$H-NMR and transverse-field (TF) $\mu$SR in a wide field region up to 9 T.

## 2. Experimental

The sample of the single crystals is prepared by a slow evaporation method [4]. The crystal is of compressed hexagonal rod shaped with a typical size of 15 mm$^3$. The $a$-axis is along the rod, and the $b$, perpendicular to the largest facet. For discussion on NMR and $\mu$SR results, we refer here the detailed crystal structure, the schematic of which is shown in Fig. 1(b). At high temperatures, the unit cell contains net two dimers [14]. This unit cell is doubled along $b$-axis at 70 K due to a structural transformation, and at lower temperatures, there are three crystallographically inequivalent $Cu_2O_6$ dimers denoted as A, B and C [3]. The $NH_4$ molecules, which are an NMR-target, locate at the interstitial sites between dimers [14]. We denote them by the number as I – IV. For example, the $NH_4$ molecule II is located closer to the dimer A compared with III or IV, indicating that we may get information concerning the inequivalent dimers through $^1$H-NMR.

$^1$H-NMR experiments were performed by the conventional spin-echo method at temperatures between 0.3 and 2.3 K, and in magnetic fields up to 8.3 T. Applied field was parallel with $b$-axis. NMR spectra were obtained by recording the spin-echo amplitude against magnetic field, which were slowly varied within a narrow region at around the zero shift position [15, 16].

Transverse-field (TF) $\mu$SR experiments were performed at the $\pi$E3 beamline of the Paul Scherrer Institute (Villigen, Switzerland), using the HAL-9500 $\mu$SR spectrometer [17]. Under the TF between 0.49 and 8.0 T parallel with $b$-axis, $\mu$SR time spectra were obtained at the temperatures of 0.1 and 1.5 K. The data length of each spectrum was 409600 bins, corresponding to the time span of 9.5 $\mu$s. They were converted to frequency spectra by FFT on the extended 933888 data points with the zero-filling technique.

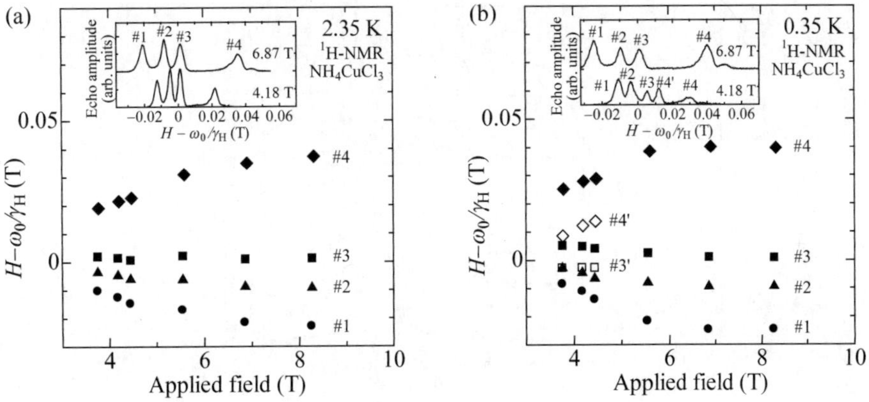


**Figure 2.** Field dependence of the peak shifts of $^1$H-NMR at (a) 2.3 K and (b) 0.3 K, corresponding to the paramagnetic and the ordered state, respectively. Typical spectral profile are shown in each inset, where $\omega_0$ is the frequency of NMR measurement. Observed peaks are denoted as #1–#4. At 0.35 K, the two of them, #3 and #4 showed splitting denoted as #3' and #4', respectively.

## 3. Results and Discussion

Typical profiles of $^1$H-NMR spectra and the field dependences of the peak shifts of $^1$H-NMR measured at 2.3 and 0.3 K are shown in Fig. 2(a) and (b), respectively. At 2.3 K, that is, in the paramagnetic state, the observed four peaks #1–#4 may come from the four $NH_4$ molecules in a unit cell. The signal from each proton nucleus within a molecule must be all integrated within a dipole width to form a single peak. Among four peaks, #4 has negative shift, and the other three, nearly zero or positive. With increasing field, the whole width of the spectrum, that is, the distance between #1 and #4 increased monotonically, and stopped increasing at $H_{C1}$. The spectral profile did not show any drastic change at $H_{C1}$. At 0.3 K, although the overall profile of the spectrum was unchanged from 2.3 K, extra splittings were observed for #3 and #4 peaks below $H_{C1}$, that is, in a slope part of the magnetization curve. With increasing field, splitting width of #3 and #3', and of #4 and #4' increased until $H = H_{C1}$, where the extra peaks #3' and #4' vanished abruptly. The temperature dependence of the spectral profile is shown in Fig. 3, where one can see that with decreasing temperature the peak #3 and #4 show a splitting at the vicinity of reported $T_N(4.45\text{ T}) = 1.08$ K.

Fourier spectra of TF-$\mu$SR taken at 0.1 K and under various fields up to 8.0 T are shown in Fig. 4. Each spectrum consisted of a central peak at the zero shift position and many satellites at the both sides of the former. The spectral shape was asymmetric; the lower side tail was longer and contained more peaks than the higher side. Although the central peak with a large amplitude may include a small contribution from the stray muons, it clearly indicates that an appreciable part of the sample is in the paramagnetic state. With increasing field, the whole width of the spectrum increased until $H = H_{C1}$, where it reached 0.2 T, and stopped further increasing. There was no drastic change in the overall spectral profile, that is, the spectrum seems to broaden with keeping the similarity of the shape, which can be clearly seen by the marker lines in Fig. 4. When the temperature was changed from 0.1 K to 1.5 K over $T_N$ under the fiield 1.0 T, an appreciable deformation took place in the spectrum including changes in peak positions as shown in the bottom of Fig. 4. In the plateau region, however, this deformation was not observed. Note that observed spectra contained more than 9 peaks, the number of which is apparently larger than that of inequivalent dimer sites, indicating that there are multiple muon

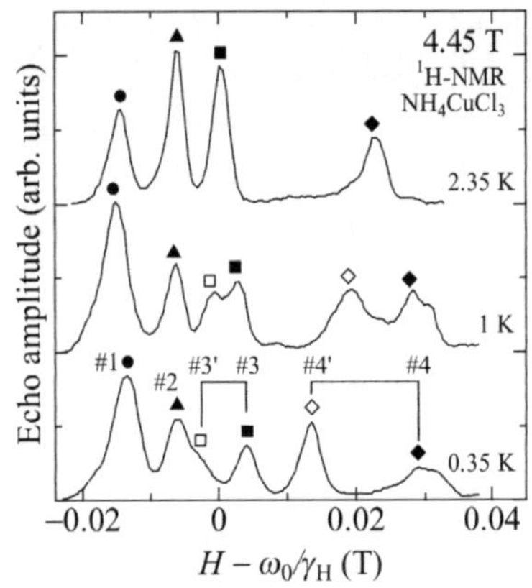


**Figure 3.** Temperature dependence of the spectra of $^1$H-NMR at 4.45 T ($H < H_{C1}$).

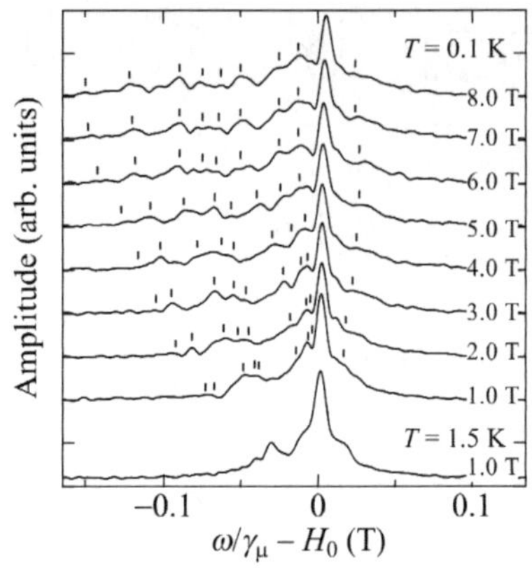


**Figure 4.** Field dependence of Fourier spectra of TF-$\mu$SR measured under constant field $H_0$ between 1.0 and 8.0 T. Vertical lines denote the peak positions.

stopping sites within a unit cell.

Now we discuss the spin state in and out of the first-plateau field region, and also over and below $T_N$. First, the observed four peaks in $^1$H-NMR spectra do indicate that the four dimers in the unit cell are magnetically inequivalent. If all the dimers are equivalently magnetized along the same direction with the same amplitude, all four $NH_4$ sites have the same hyperfine field, hence one single peak in $^1$H-NMR spectrum. Note that this dimer inequivalency exists not only in the ordered state but also in the paramagnetic state. These results are consistent with our previous $\mu$SR report [9]. Next, we discuss the relation between the magnetization process and the peak shifts. The peak shifts in spectra of both $^1$H-NMR and TF-$\mu$SR increased monotonically with increasing field and stopped at $H_{C1} = 5$ T. This behavior simply traces the development of uniform magnetization $M_z$ with increasing applied field [4]. Therefore, the difference in the shift of four peaks is considered to come from that in the $z$-component of the local magnetization borne by inequivalent dimers. For a rough estimation, if one assumes that only the A dimer is fully polarized within the plateau field region as proposed by Matsumoto [7], we can obtain the hyperfine field at $NH_4$ site II as $-350$ Oe, which is in accordance with the observation. The fact that the spectral profile was unchanged at $H_{C1}$ demonstrates the continuity in the spin state with increasing the field. In other words, there is no solidification of field-induced magnons on entering the plateau region [18]. The detailed assignment of the observed NMR peaks to each $NH_4$ site is in progress.

In the field of slope region, the peaks #3 and #4 split at low temperature below $T_N$, supporting the existence of magnetic order with the ordered moment $M_{xy}$ [7, 12]. The significant change in $\mu$SR spectra above and below $T_N$ at 1 T is also considered to be due to this effect. In the plateau region, $T_N$ becomes zero, and hence both the NMR and $\mu$SR spectrum should be unchanged with temperatures, as was observed.

Finally, we mention the possibility of lattice deformation. As stated above, in $TlCuCl_3$ isomorphic to the present system, the magnetic order was accompanied by the sizable change in the lattice constants and hence the finite jump in the hyperfine field at $T_N$ [13]. In $NH_4CuCl_3$, we have observed the splitting width of the peaks #3 and #4 abruptly vanishes as entering the plateau region. However, in order to judge whether or not transition at $H_{C1}$ is of first order, more detailed measurements in the vicinity of the phase boundary are necessary, and are now in progress.

## 4. Summary

We have investigated the spin states of the $S = 1/2$ three dimensional spin dimer system $NH_4CuCl_3$ by high-field TF-$\mu$SR and $^1$H-NMR. We have observed multiple peaks with different shifts both in NMR and $\mu$SR spectra, which clearly demonstrate the existence of inequivalent dimers in a unit cell. No drastic change in the spectral profile was observed at $H_{C1}$, indicating that the spin state changes continuously over $H_{C1}$, that is, no liquid-solid transition takes place in the field-induced magnons.

## 5. Acknowledgments

This work was partly supported by JSPS KAKENHI Grant Number 15K05148 and 15J05303.

## References

[1] Rice T M 2002 *Science* **298** 760–761 ISSN 0036-8075 (*Preprint* http://science.sciencemag.org/content/298/5594/760.full.pdf) URL http://science.sciencemag.org/content/298/5594/760
[2] Kurniawan B, Ishikawa M, Kato T, Tanaka H, Takizawa K and Goto T 1999 *Journal of Physics: Condensed Matter* **11** 9073 URL http://stacks.iop.org/0953-8984/11/i=46/a=309
[3] Rüegg C, Oettli M, Schefer J, Zaharko O, Furrer A, Tanaka H, Krämer K W, Güdel H U, Vorderwisch P, Habicht K, Polinski T and Meissner M 2004 *Phys. Rev. Lett.* **93**(3) 037207 URL http://link.aps.org/doi/10.1103/PhysRevLett.93.037207
[4] Shiramura W, Takatsu K I, Kurniawan B, Tanaka H, Uekusa H, Ohashi Y, Takizawa K, Mitamura H and Goto T 1998 *Journal of the Physical Society of Japan* **67** 1548–1551 (*Preprint* http://dx.doi.org/10.1143/JPSJ.67.1548) URL http://dx.doi.org/10.1143/JPSJ.67.1548
[5] Kolezhuk A K 1999 *Phys. Rev. B* **59**(6) 4181–4188 URL http://link.aps.org/doi/10.1103/PhysRevB.59.4181
[6] Oshikawa M, Yamanaka M and Affleck I 1997 *Phys. Rev. Lett.* **78**(10) 1984–1987 URL http://link.aps.org/doi/10.1103/PhysRevLett.78.1984
[7] Matsumoto M 2015 *Journal of the Physical Society of Japan* **84** 034701 (*Preprint* http://dx.doi.org/10.7566/JPSJ.84.034701) URL http://dx.doi.org/10.7566/JPSJ.84.034701
[8] Inoue H, Tani S, Hosoya S, Inokuchi K, Fujiwara T, Saito T, Suzuki T, Oosawa A, Goto T, Fujisawa M, Tanaka H, Sasaki T, Awaji S, Watanabe K and Kobayashi N 2009 *Phys. Rev. B* **79**(17) 174418 URL http://link.aps.org/doi/10.1103/PhysRevB.79.174418
[9] Matsui K, Oosawa A, Yoshizawa K, Goto T, Watanabe I, Suzuki T, Fujisawa M, Tanaka H, Biswas P K and Amato A 2014 *Journal of Physics: Conference Series* **551** 012007 URL http://stacks.iop.org/1742-6596/551/i=1/a=012007
[10] Tanaka H, Takatsu K I, Shiramura W and Ono T 1996 *Journal of the Physical Society of Japan* **65** 1945–1948 (*Preprint* http://dx.doi.org/10.1143/JPSJ.65.1945) URL http://dx.doi.org/10.1143/JPSJ.65.1945
[11] Nikuni T, Oshikawa M, Oosawa A and Tanaka H 2000 *Phys. Rev. Lett.* **84**(25) 5868–5871 URL http://link.aps.org/doi/10.1103/PhysRevLett.84.5868
[12] Matsumoto M, Normand B, Rice T M and Sigrist M 2002 *Phys. Rev. Lett.* **89**(7) 077203 URL http://link.aps.org/doi/10.1103/PhysRevLett.89.077203
[13] Vyaselev O, Takigawa M, Vasiliev A, Oosawa A and Tanaka H 2004 *Phys. Rev. Lett.* **92**(20) 207202 URL http://link.aps.org/doi/10.1103/PhysRevLett.92.207202
[14] Willett R D, Dwiggins C, Kruh R F and Rundle R E 1963 *The Journal of Chemical Physics* **38** 2429–2436 URL http://scitation.aip.org/content/aip/journal/jcp/38/10/10.1063/1.1733520
[15] Matsui K, Fujisawa M, Hagiwara K, Hoshino Y, Goto T, Sasaki T, Tanaka H, Okubo S and Ohta H *Ground State of the Spin-1/2 Chain of Green Dioptase at High Fields* (*Preprint* http://journals.jps.jp/doi/pdf/10.7566/JPSCP.3.014011) URL http://journals.jps.jp/doi/abs/10.7566/JPSCP.3.014011
[16] Kanada K, Saito T, Oosawa A, Goto T and Suzuki T 2007 *Journal of the Physical Society of Japan* **76** 064706 (*Preprint* http://dx.doi.org/10.1143/JPSJ.76.064706) URL http://dx.doi.org/10.1143/JPSJ.76.064706
[17] Goto T, Suzuki T, Kanada K, Saito T, Oosawa A, Watanabe I and Manaka H 2008 *Phys. Rev. B* **78**(5) 054422 URL http://link.aps.org/doi/10.1103/PhysRevB.78.054422

[18] Takigawa M, Kodama K, Horvatić M, Berthier C, Matsubara S, Kageyama H, Ueda Y, Miyahara S and Mila F 2006 *Journal of Physics: Conference Series* **51** 23 URL http://stacks.iop.org/1742-6596/51/i=1/a=004

IOP Conf. Series: Journal of Physics: Conf. Series **828** (2017) 012012 doi:10.1088/1742-6596/828/1/012012

# Possible low spin state of Fe ions in $HoBaFe_4O_7$

**K. Kamazawa and T. Moyhoshi**

Comprehensive Research Organization for Science and Society (CROSS), Tokai, Ibaraki 319-1106, Japan

E-mail: k_kamazawa@cross.or.jp

**Abstract.** Effective magnetic moment of $YBaFe_4O_{7-\delta}$ (YBFO) and $HoBaZn_3GaO_7$ is obtained as 4.88 $\mu_B$ (/magnetic atom) and 10 $\mu_B$ (/magnetic atom), respectively from magnetic susceptibility. Nevertheless that of $HoBaFe_4O_7$ (HBFO) is almost half magnitude (2.77 $\mu_B$ /magnetic atom) than simply summed up the former two effective moments, although magnetic interaction between Ho and Fe ions is absent. The small effective moment of HBFO suggests different spin state in comparison with YBFO. Taking consideration of our recent neutron inelastic scattering result of HBFO, the small effective moment of HBFO is considered by low spin state of Fe ions. Low spin state is hardly expected in cubic system of Fe ions usually, but low spin state is considered to be realized in HBFO due to spin multiple splitting by dynamical Jahn-Teller effect.

## 1. Introduction

A cubic antiferromagnet of $RBaFe_4O_7$ (RBFO) (R = Y and Ho) has a network of corner-sharing-tetrahedra and spin frustrated features can be expected, as well as spinel[1]-[5] and pyrochlore[6][7] system.(See Figure 1) Unlike spinel and pyrochlore, however, the Fe ions of corner-sharing-tetrahedra have mixed valence state $Fe^{2+}$ and $Fe^{3+}$ with a number ratio of 3 : 1. When the ionic radius of R is small, the crystal structure tends to have cubic symmetry [8][9](, although there are some reports about lattice distortion.[10][11]) Even though $Fe^{2+}$ in tetrahedral symmetry is a Jahn-Teller active ion, both R = Y and Ho systems do not show any charge ordering[12]-[14] and static lattice distortion to the lowest temperature that we measured to 4K.

In our recent inelastic neutron scattering investigations of RBFO (R = Y[15] and Ho[16]) powder samples, characteristic magnetic excitations were revealed in the energy($\omega$)-wave vector($Q$) space, which are associated with geometrical spin frustration. The $YBaFe_4O_{7-\delta}$ (YBFO) system shows inelastic magnetic scattering extending up to rather high $\omega$ values (~80 meV) at around $Q = Q_p$(~1.25 Å$^{-1}$), the wave vector corresponding to a hexagon correlation of Fe-spins. We thus interpret the streak scattering along the $\omega$-axis at $Q_p$ indicating the creations and annihilations of hexagon spin clusters - that is, the spins are fluctuating in wide energy range but maintaining antiferromagnetic hexagon correlation. On the other hand, in $HoBaFe_4O_7$ (HBFO), dispersionless modes are dominantly observed at several discrete energy levels. Some energy levels cannot be attributed to crystal field levels of $Ho^{3+}$ in octahedral symmetry. Some levels are rather explained by vibronic state of $Fe^{2+}$ dynamical Jahn-Teller effect with spin-orbit coupling. That is, spin-orbital-lattice degree of freedom is coupled to each other in HBFO. The difference of R ion affects the dynamics of the frustrated system RBFO, resulting in rich

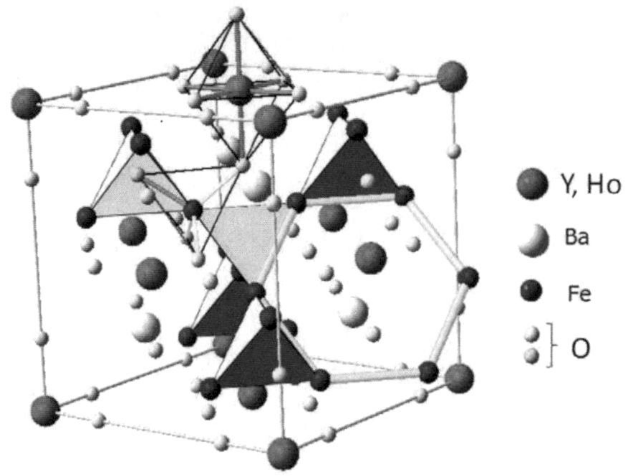


**Figure 1.** (color online) (a) Crystal structure of cubic 114 ferrite $RBaFe_4O_7$ (R=Y and Ho). The number ratio of $Fe^{2+}$ and $Fe^{3+}$ is 3:1. They form corner-sharing-tetrahedra. The Fe ions are on tetrahedral sites and the R ion is on octahedral sites. The (green) bold sticks indicate one of the hexagons of Fe spins.

varieties of characteristic excitation such as strong spin fluctuations over wide energy range at characteristic wave vectors or dynamical interaction among spin-orbital-lattice degrees of freedom as dispersionless modes. In the magnetic susceptibility measurements of RBFO (R =Y and Ho), we obtained 4.88 $\mu_B$ (/magnetic atom) effective moment of YBFO[15] that indicates high spin state of $Fe^{2+}$ and $Fe^{3+}$. On the other hand, only half magnitude of the effective moment (2.77 $\mu_B$ /magnetic atom) was obtained for HBFO[15]. The small effective moment of HBFO suggests different spin state in contrast to YBFO. This time, we have synthesized $HoBaZn_3GaO_7$ and observed the effective moment from the magnetic susceptibility, then taking into consideration of the results of spin multiple splitting of HBFO[16], we have tried to explain the origin of the small effective moment of HBFO. The crystal structure of $HoBaZn_3GaO_7$ is not isostructural with cubic RBFO (R = Y and Ho), but the differences of crystal structure can be ignored in the estimation of effective moment, here).

## 2. Sample Preparation and Experimental Method

A powder sample of $HoBaZn_3GaO_7$ was prepared by a two-step solid-state reaction. First, a mixture of $BaCO_3$ and $Ga_2O_3$ with a molar ratio of 1 : 1 was ground and pressed into pellets, which were heated at 1200 °C for 12 hours in air to obtain $BaGa_2O_4$. Next, a stoichiometric mixture of $Ho_2O_3$, $BaGa_2O_4$, and ZnO was ground and pressed into pellets, which were heated at 1100 °C for 24 hours in argon gas. Magnetic susceptibility measurements were performed with a SQUID (Quantum Design) system. The preparation of YBFO and HBFO are described somewhere in Ref. [15] and [16].

## 3. Results and Discussion

Figure 2 shows field-cooling magnetic susceptibility and inverse magnetic susceptibility of $HoBaZn_3GaO_7$, YBFO, and HBFO. By Curie-Weiss fitting at the temperature region about 100 K $\sim$ 300 K, we obtained Curie constant ($C$) as $C_{HoBaZn3GaO7} = 13.26$ emu·K/mol (effective moment is 10.29 $\mu_B$; 10.29 $\mu_B$ /magnetic atom) for $HoBaZn_3GaO_7$, $C_{YBFO} = 47.63$ emu·K/mol (effective moment is 19.52 $\mu_B$; 4.88 $\mu_B$ /magnetic atom) for YBFO[15], and $C_{HBFO} = 23.98$ emu·K/mol (effective moment is 13.85 $\mu_B$; 2.77 $\mu_B$ /magnetic atom) for HBFO[16]. If all spin

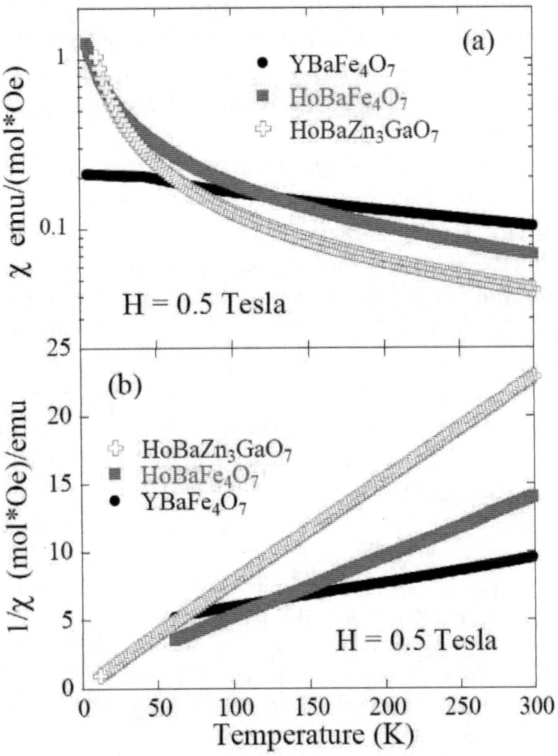


**Figure 2.** (color online) (a) Field-cooling magnetic susceptibility under the magnetic field (H) of 0.5 Tesla. (b) Field-cooling the inverse magnetic susceptibility under the magnetic field (H) of 0.5 Tesla.

state are the same, effective moment of HBFO should be obtained as almost summed value of $HoBaZn_3GaO_7$ and YBFO. When we refer to another frustrated magnets, *e.g.* $R_2Mo_2O_7$ (R = Y and Gd) pyrochlore system, the effective moments are observed 1.7 $\mu_B$ (/magnetic atom) in $Y_2Mo_2O_7$[17], 7.94 $\mu_B$ (/magnetic atom) in $Gd_2Ti_2O_7$[18][19], and 7.7 $\mu_B$ (/magnetic atom) in in $Gd_2Mo_2O_7$[17]. The effective moment is obtained almost summed value of $Y_2Mo_2O_7$ and $Gd_2Ti_2O_7$, as we expected. A little reduction (~80 % of actual summed value) would be due to quantum fluctuations and/or due to spin-orbit coupling by ionic combination effect.[20] Only half magnitude of effective moment of 13.85 $\mu_B$ is observed in our HBFO case, although $Ho^{3+}$ is 10 $\mu_B$, $Fe^{2+}$ is 4 $\mu_B$, and $Fe^{3+}$ is 5 $\mu_B$, resulting 10 $\mu_B$ +(3× 4 $\mu_B$)+(1× 5 $\mu_B$) = 27 $\mu_B$ should be observed in total. Here, it is hard to attribute the reduced effective moment to $Ho^{3+}$ in HBFO, because there is no positive reason for the $Ho^{3+}$ moment reduction. Then, possible origins are considered to be attributed from the Fe ions.

In 3*d* transition metal oxide, crystal field splitting energy is larger than spin-orbit coupling energy and electrons with the parallel spin try to set different degenerate level first due to the Pauli-exclusion-principle. Once it became full, electrons with antiparallel spin start to set the same degenerate level. This is known as Hunt's rule. Therefore high spin state is often taken by 3*d* transition metal. In fact, the effective moment of YBFO suggests high spin state of Fe ions. When crystal field splitting is larger than Hunt's coupling energy, low spin state can be realized.

This situation is sometimes released by lattice distortion etc, but there is no lattice distortion in HBFO. Let's remind spin multiple splitting of HBFO as vibronic mode due to dynamical Jahn-Teller effect with spin-orbit coupling without the crystal structure distortion.[16] Even though the splitting energy level between $T_{2g}$ and $E_g$ is not enough to realize low spin state, since the degeneracy of $E_g$ level is lifted, the electrons in $T_d$ symmetry of $Fe^{2+}$ and $Fe^{3+}$ are not affected by Hunt's effect, and is considered to achieve low spin state. When Fe ions takes low spin state the spin value $s$ of $Fe^{3+}$ is $s = 1/2$ and that of $Fe^{3+}$ is $s = 1$. In this case, the effective magnetic moment becomes 3.4 $\mu_B$ (/magnetic atom) = (10 $\mu_B$ + 2 $\mu_B$ + 2 $\mu_B$ + 2 $\mu_B$ + 1 $\mu_B$) / 5 magnetic atoms, in addition to the 10 $\mu_B$ of $Ho^{3+}$ moment. With considering a reduction factor, *i.e.* ~80 %, it becomes ~2.72 $\mu_B$ and almost the same value with the experimental one. Although $Fe^{3+}$ is $S^{5/2}$ state and is not necessary to follow the same way of $Fe^{2+}$, we have treated $Fe^{3+}$ as supposing in the multiple splitting levels. Because, $Fe^{2+}$ and $Fe^{3+}$ form corner-sharing-tetrahedra with the ratio of 3:1, and $Fe^{3+}$ is also considered to be affected by multiple splitting through the spin-orbit-lattice interaction by dynamical Jahn-Teller effect.

In conclusion, low spin state is realized in $HoBaFe_4O_7$ due to vibronic state by dynamical Jahn-Teller effect. In other words, without lattice distortion, observed the small effective magnetic moment is a strong evidence of spin multiple splitting state of $HoBaFe_4O_7$.

## References

[1] Kamazawa K, Tsunoda Y, Odaka K and Kohn K 1999 *J. Phys. Chem. Sol.* **60** 1261
[2] Lee S -H, Broholm C, Kim T H, Ratcliff II W and Cheong S -W 2000 *Phys. Rev. Lett.* **84** 3718
[3] Lee S -H, Broholm C, Ratcliff W, Gasparovic G, Huang Q, Kim T H and Cheong S -W 2002 *Nature* **418** 856
[4] Kamazawa K, Tsunoda Y, Kadowaki H and Kohn K 2003 *Phys. Rev.* B **68** 024412
[5] Kamazawa K, Park S, Lee S -H, Sato T J and Tsunoda Y 2004 *Phys. Rev.* B **70** 024418, 2010 *Phys. Rev.* B **81** 099906
[6] Greedan John E 2006 *Journal of Alloys and Compounds* **408-412** 444
[7] Gardner Jason S, Gingras Michel J P and Greedan John E 2010 *Rev. Mod. Phys.***82** 53
[8] Pralong V, Caignaert V, Maignan A and Raveau B 2009 *J. Mater. Chem.* **19** 8335
[9] Caignaert V, Abakumov A M, Pelloquin D, Pralong V, Maignan A, Tendeloo G Vanand Raveau B 2009 *Chem. Mater.* **21** 1116-1122
[10] Duffort V, Caignaert V, Pralong V, Barrier N, Raveau B, Avdeev M, Zheng H and Mitchell J F 2012 *J. Solid. State. Chem.* **191** 225
[11] Duffort V, Sarkar T, Caignaert V, Pralong V, Raveau B, Avdeev M, Cervellino A, Waerenborgh J C and Tsipis E V 2013 *J. Sol. Stat. Chem.* **205** 225
[12] Verwey E J W and Haayman P W 1941 *Physica* **8** 979
[13] Verwey E J W, Hayman P W and Romeijn C W 1947 *J. Chem. Phys.* **15** 181
[14] Walz F 2002 *J. Phys. Cond. Mat.* **14** R285
[15] Kamazawa Kazuya, Ishikado Motoyuki,Ohira-Kawamura Seiko, Kakurai Kazuhisa, Nakajima Kenji, Kawakita Yukinobu, Yamada Kazuyoshi, Arai Masatoshi and Sato Masatoshi 2015 *J. Phys. Soc. Jpn.* **84** 104711
[16] Kamazawa K, Ishikado M, Ohira-Kawamura S, Kawakita Y, Kakurai K, Nakajima K, Sato M 2016 To be submitted.
[17] Raju N P, Gmelin E and Kremer R K 1992 *Phys. Rev.* B **46** 5405
[18] Raju N P, Dion M, Gingras M J P, Mason T E and Greedan J E 1999 *Phys. Rev.* B **59** 14489
[19] Luo Gang, Hess Steven T, Corruccini L R 2001 *Phys. Lett.* A **291** 306
[20] Schmitz Robert, Entin-Wohlman Ora, Aharony Amnon, Harris A. Brooks and M?ller-Hartmann Erwin 2005 *Phys. Rev.* B **71** 144412

# Physical properties in the cluster-based magnetic-diluted triangular lattice antiferromagnets $Li_2Sc_{1-x}Sn_xMo_3O_8$

**Yuya Haraguchi[1], Chishiro Michioka[1], Hiroaki Ueda[1], Akira Matsuo[2], Koichi Kindo[2], Kazuyoshi Yoshimura[1,3,4]**

[1] Department of Chemistry, Graduate School of Science, Kyoto University, Kyoto 606-8502, Japan
[2] The Institute for Solid State Physics, The University of Tokyo, 5-1-5, Kashiwanoha, Kashiwa, Chiba 277-8581, Japan
[3] Research Center for Low Temperature and Materials Sciences, Kyoto University, Kyoto 606-8501, Japan
[4] Institute for Liberal Arts and Sciences, Kyoto University, Kyoto 606-8501, Japan

E-mail: chiyuya@kuchem.kyoto-u.ac.jp, kyhv@kuchem.kyoto-u.ac.jp

**Abstract.** We have investigated the physical properties of the magnetic diluted triangular lattice antiferromagnetic system $Li_2Sc_{1-x}Sn_xMo_3O_8$. For all compounds, no mangetic ordering has been observed. On the other hand, the partial spin disappearing behavior is found in all Sn-substituted compounds except $x = 0$, which has been also observed in the similar magnetic system $LiZn_2Mo_3O_8$. Considering the relationship between the crystal structure and the magnetism, the partial spin disappearance is properly explained by a formation of the valence bond glass derived from the randomness effect

## 1. Introduction

Geometrical frustrated magnetism have attracted much attention because of their possibility to host novel phenomena. In particular, quantum spin frustrated magnets are expected to exhibit exotic ground states like a quantum spin liquid (QSL) state [1, 2, 3]. A strong spin frustration is realized in a characteristic lattice such as triangular, kagome (in 2D), face-centered cubic and pyrochlore lattice (in 3D). In realistic compounds, the highly frustrated spin liquid state is hard to be realized owing to the higher order magnetic interactions or coupling to other degrees of freedom such as lattice, valence, and orbital instabilities. Newly produced ground states may contain novel physics depending on characteristics of each compound.

Recently, we discovered $S = 1/2$ triangular lattice cluster antiferromagnet $Li_2ScMo_3O_8$ [4]. In this compound, $Mo_3$ trimers form a triangular lattice, and the $Mo_3$ layers and nonmagnetic cations are stacked alternately as shown in the Fig.1 (a) and (b). The magnetic state of $[Mo_3]^{11+}$ in $Li_2ScMo_3O_8$ is $S = 1/2$ as shown in Fig. 1(c). Then, $Li_2ScMo_3O_8$ is categorized as the $S = 1/2$ triangular lattice antiferromagnets. In our previous investigations, it has been found that $Li_2ScMo_3O_8$ is a powerful candidate of the QSL. In the similar magnetic systems, the isostructural $Li_2InMo_3O_8$ exhibits a conventional magnetic ordering with 120° spin configuration [4], and the other magnetic system $LiZn_2Mo_3O_8$ shows a formation of the valence bond solid on the honeycomb sublattice without magnetic ordering [5]. Thus, the $Mo_3$ trimer-based triangular lattice antiferromagnets exhibit various phenomena based on a frustrate magnetism. Therefore,

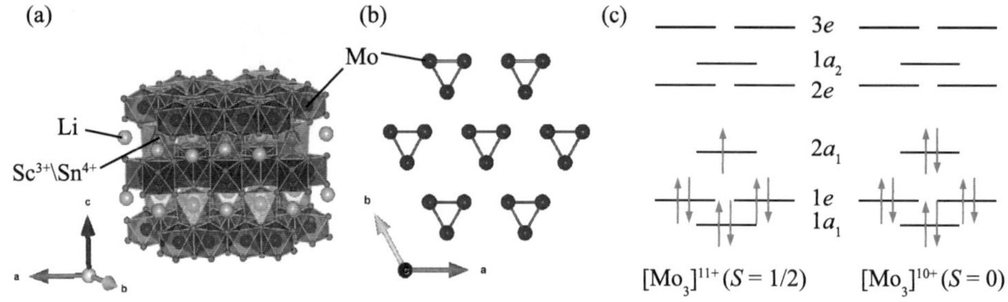


**Figure 1.** (Color online) (a) The crystal structure of $Li_2Sc_{1-x}Sn_xMo_3O_8$. (b) $Mo_3$ trimers forming a triangular lattice. (c) Schematic 4*d* orbital energy level of the $[Mo_3]^{11+}/[Mo_3]^{10+}$ trimer with the electron configurations in $Li_2Sc_{1-x}Sn_xMo_3O_8$.

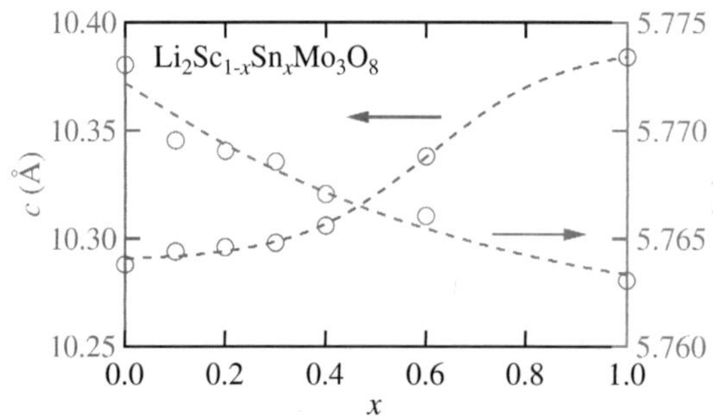


**Figure 2.** (Color online) The trigonal lattice parameters $a$ and $c$ in $Li_2Sc_{1-x}Sn_xMo_3O_8$ as a function of $x$. The dashed lines are a guide to the eye.

it is necessary to reveal the different ground state among $Mo_3$ cluster compounds, which will help understanding the QSL behavior in $Li_2ScMo_3O_8$.

In this work, we report on the physical properties of the magnetic diluted $Li_2Sc_{1-x}Sn_xMo_3O_8$ system. The Sn-substitution reduces the valence of Mo, which reflects the magnetic dilution to the $S = 1/2$ triangular lattice system by the nonmagnetic $[Mo_3]^{10+}$ trimers ($S = 0$) instead of the $[Mo_3]^{11+}$ trimers ($S = 1/2$). That is, the chemical formula can be written as $Li_2Sc_{1-x}Sn_x[Mo_3]^{11+}{}_{1-x}[Mo_3]^{10+}{}_xO_8$. For all Sn-substituted compounds without $x = 0$, the partial spin disappearing behavior is found similar to $LiZn_2Mo_3O_8$. The relationship between the magnetic properties and the crystallographic structure for both systems indicates that the randomness is the possible origin to emerge the partial spin disappearance caused by the valence bond glass formation.

## 2. Experimental Methods

Polycrystalline samples of $Li_2Sc_{1-x}Sn_xMo_3O_8$ were prepared using conventional solid-state reactions. These samples were characterized by powder x-ray diffraction (XRD) on a diffractometer with Cu$K\alpha$ radiation. As shown in Fig. 2, the lattice parameters $a$ and $c$ in $Li_2Sc_{1-x}Sn_xMo_3O_8$ monotonically vary with increasing $x$, suggesting the successful solid solution. The temperature dependence of the magnetization was measured under several magnetic fields up to 7 T by using a magnetic property measurement system (MPMS; Quantum

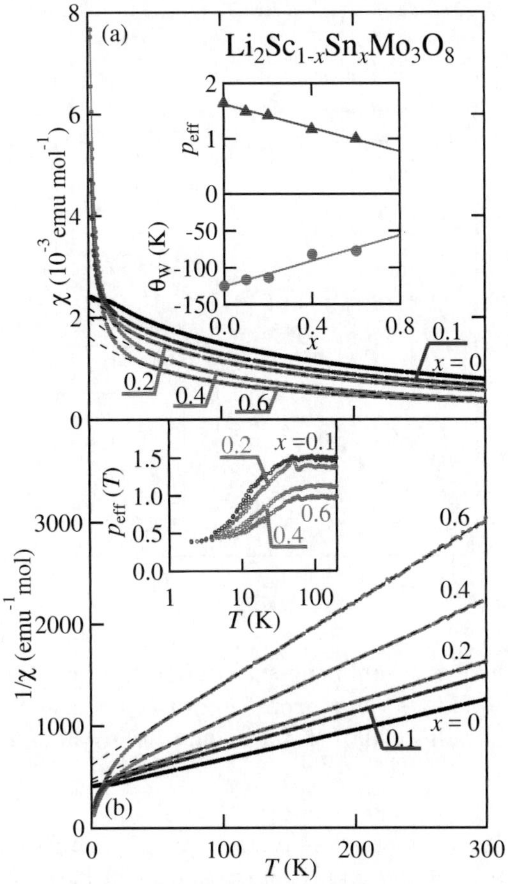


**Figure 3.** (Color online) (a) The temperature dependence of the magnetic susceptibility $\chi$ in $Li_2Sc_{1-x}Sn_xMo_3O_8$. The inset shows the composition dependence of the the effective paramagnetic Bohr magneton $p_{\rm eff}$ and the Curie-Weiss temperatures at the high-temperature region. (b) The $1/\chi$ vs. $T$ plot. The inset shows the temperature dependence of the normalized effective paramagnetic Bohr magneton $p_{\rm eff}(T)$.

Design) equipped at the LTM Research Center, Kyoto University. Magnetization measurements up to 60 T were performed using an induction method with a multilayer pulse magnet at the Ultra-high Magnetic Field Laboratory of the Institute for Solid State Physics at the University of Tokyo.

## 3. Results and Discussion

Figure 3(a) shows the temperature dependence of the magnetic susceptibility $\chi$ in $Li_2Sc_{1-x}Sn_xMo_3O_8$. For all compositions, they obey in the Curie-Weiss law in the high temperature region and show no anomaly accompanied by a magnetic ordering or a spin glass freezing. The Weiss temperature $\theta_W$ and the effective paramagnetic Bohr magneton $p_{\rm eff}$ monotonically vary with increasing Sn substitution $x$, which is plotted in the inset of the Fig. 3(a). The reduction of $p_{\rm eff}$ by increasing $x$ suggests the increase of the nonmagnetic $[Mo_3]^{10+}$ trimers instead of the $S = 1/2$ $[Mo_3]^{11+}$ trimers. The reduction of $\theta_W$ is originated in the

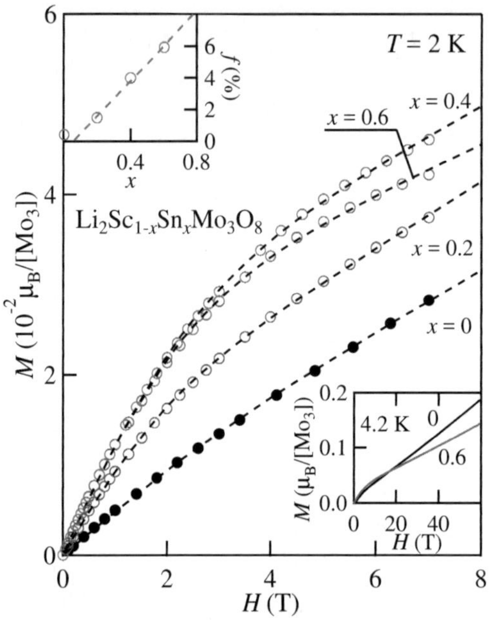


**Figure 4.** (Color online) The isothermal magnetization at 2 K (circles) and those fitting curves with Eq. (1). The upper inset shows the percentage of free spins. The downer inset shows the isothermal magnetization up to $H = 60$ T at 4.2 K only for $x = 0$ and 0.6.

reduction of the number of the magnetic couplings from the magnetic dilution.

For all composition except $x = 0$, $\chi$ shows an enhancement and a deviation from the Curie-Weiss law in the low temperature region. In addition, as shown in Fig. 3(b), a change in the slope of $1/\chi$ as a function of temperature is observed at low temperatures, suggesting the partial disappearance of the paramagnetic spins. The inset of Fig. 3(b) shows the temperature dependence of $p_{\rm eff}$, estimated using the relation,

$$p_{\rm eff} \simeq \sqrt{\frac{8}{d\chi^{-1}/dT}}. \qquad (1)$$

For all Sn-substituted compounds, $p_{\rm eff}$ gradually reduces with decreasing temperature. This behavior is very similar to the similar magnetic compound $LiZn_2Mo_3O_8$, in which the partial disappearance of the paramagnetic spins has been explained by the construction of a valence bond solid on the honeycomb sublattice. The the partial spin-disappearance in the $Li_2Sc_{1-x}Sn_xMo_3O_8$ system is also explained by a formation of partial spin singlet. The finite $p_{\rm eff}$ value in low temperatures means the presence of the free spins which have been orphaned by the partial spin singlet formation.

Figure 4 shows the isothermal magnetization measured in the field up to $H = 7$ T at $T = 2$ K. For all Sn-substituted compounds, $M$-$H$ curves show the nonlinear behavior. The deviation from the linear relationship $M \propto H$ becomes larger with increasing $x$, indicating that the contribution of free spins becomes large with increasing $x$. In order to estimate the amount of free spins, we analyze the magnetization within the framework of the model that $M$ is contributed by two components from the singlet-formed spins and the nearly free spins,

$$M(H,T) = \chi(T)H + N_A\mu_B f S_f g_f B_s(g_f S_f H\mu_B/k_B(T-\theta_f)), \qquad (2)$$

**Table 1.** Results of fits to the isothermal magnetization data using the formula described in text. The parameters $g_f$ and $S_f$ are fixed to 2 and 0.5, for respectively.

| $x$ | $f$(%) | $\theta_f$(K) | $g_f$ | $S_f$ | $\chi$ (emu/mol-$[Mo_3]^{11+}$) |
|---|---|---|---|---|---|
| 0 | 0.43(1) | 0.020(2) | 2 | 0.5 | 0.00191(3) |
| 0.2 | 1.49(2) | 0.65(2) | 2 | 0.5 | 0.00258(3) |
| 0.4 | 4.00(5) | 0.26(2) | 2 | 0.5 | 0.00301(7) |
| 0.6 | 5.93(6) | 0.38(3) | 2 | 0.5 | 0.00381(9) |

where $\chi$ represents the susceptibility derived from the singlet-formed spins, and $S_f$, $g_f$, and $\theta_f$ denote, the Lande $g$-factor, spin quantum number, and Weiss temperature for early free spins. The symbols $H$, $N_A$, $k_B$, $\mu_B$ and $B_s$ represent the magnetic field, the Avogadro number, the Boltzman constant, the Bohr magneton, and the Brillouin function, respectively. The parameter $f$ represents the percentage of free spins against the magnetic $[Mo_3]^{11+}$ trimers. The fitting results are listed in Table 1. The estimated values of $\theta_f$ are very small, indicating almost no correlation among the free spins. The upper inset shows the $x$ dependence of the estimated $f$ value. For all Sn-substituted compounds, the contribution of the free spin to the magnetization process is relatively larger than that of the non-substituted compound. Considering that the $f$ value is relatively small in $x = 0$ and remarkably increases with increasing $x$, the main part of the free spin in the low temperature phase would be derived from the randomness effect. Note that the amount of free spins should be due to extrinsic factors, for example, defects and impurity spins because of the finite value of $f$ in $x = 0$. The magnetization process up to 60 T only for $x = 0$ and 0.6 are shown in the downer inset of Fig. 4. Neither saturated behavior nor field-induced transition is observed and each $M$ shows linear behavior against $H$ in the high magnetic field region, indicating that the spin singlet is not broken up to 60 T.

For $Li_2Sc_{1-x}Sn_xMo_3O_8$ and $LiZn_2Mo_3O_8$, the similar $p_{\rm eff}$ reduction has been observed. However, such the reduction is observed only in the system with the crystallographic disorder. In the case of $LiZn_2Mo_3O_8$, previous studies have clarified the presence of the mixture of Li and Zn ions [6]. And in the case of $Li_2Sc_{1-x}Sn_xMo_3O_8$, the intensional randomness is present. Thus, it is reasonable to think that the partial spin singlet formation to make the reduction of $p_{\rm eff}$ is caused by the randomness. Such a behavior is also observed in the triangular lattice $Sc_2Ga_2CuO_7$ [8] and $Ba_3IrTi_2O_9$ [9], and the kagome lattice Herbertsmithite $ZnCu_3(OH)_6Cl$ [10] and $ZnCu_3(OH)_6SO_4$ [11]. They also have crystallographic disorders of magnetic/nonmagnetic ions on the frustrated lattice. In the theoretical prediction, the randomness-induced valence bond glass model can explain the partial disappearance of paramagnetic spins [7]. In this model, the appearance of a small fraction of free spins to produce the Curie susceptibility is also predicted; leave some unpaired free spins would be orphaned by the random dimer spin pairing. Therefore, the origin of the the valence bond glass behavior observed in the $Mo_3$ cluster systems would be the randomness effect derived from the magnetic dilution. And such randomness effect to emerge the valence bond glass behavior would be characteristic in the highly frustrated magnets.

## 4. Summary

We studied the physical properties of the magnetic diluted triangular lattice antiferromagnetic systems $Li_2Sc_{1-x}Sn_xMo_3O_8$. With increasing of Sn substitution $x$, the nonmagnetic $[Mo_3]^{10+}$ clusters increase instead of the $S = 1/2$ $[Mo_3]^{11+}$ clusters, which results in the introduction of a magnetic dilution to the magnetic systems on the triangular lattice. For Sn substituted compounds, the partial disappearance of paramagnetic spins without magnetic orderings has been observed. This behavior is also observed in the similar magnetic system $LiZn_2Mo_3O_8$. Considering the relationship between the magnetic properties and the crystallographic structure for both systems, the randomness effect is the possible driving-force to emerge the partial disappearance of paramagnetic spins caused by the valence bond glass formation. Furthermore, it is concluded that the quantum spin liquid behavior observed in $Li_2ScMo_3O_8$ is an essential character on the triangular lattice antiferromagnets without a randomness effect.

**Acknowledgments**
This work is supported by Grant-in-Aids for Scientific Research from MEXT of Japan [Grants No. 16H04131 and No. 26410089]. Y. H. is supported by Research Fellow of Japan Society for the Promotion of Science [Grants No. 16J04048].

## References

[1] Anderson P W 1973 *Mat. Res. Bull.* **8**, 153
[2] Moessner R and Sondhi S L 2002 *Prog. Theor. Phys. Suppl.* **145**, 37
[3] Lee P A 2008 *Science* **321**, 1306
[4] Haraguchi Y, Michioka C, Imai M, Ueda H and Yoshimura K 2015 *Phys. Rev. B* **92**, 014409
[5] Sheckelton J P, Neilson J R, Soltan D G and McQueen T M 2012 *Nat. Mater.* **11**, 493
[6] Torardi C C and McCarley R E 1985 *Inorg. Chem.* **24**, 476
[7] Watanabe K, Kawamura H, Nakano H, Sakai T 2014 *J. Phys. Soc. Jpn.* **83**, 034714
[8] Kumar R, Khuntia P, Sheptyakov D, Freeman P G, Rønnow H M, Koteswararao B, Baenitz M, Jeong M and Mahajan A V 2015 *Phys. Rev. B* **92**, 180411(R)
[9] Dey T, Mahajan A V, Khuntia P, Baenitz M, Koteswararao B and Chou F C 2012 *Phys. Rev. B* **86**, 140405
[10] Helton J S, Matan K, Shores M P, Nytko E A, Bartlett B M, Yoshida Y, Takano Y, Suslov A, Qiu Y, Chung J -H, Nocera D G and Lee Y S 2007 *Phys. Rev. Lett.* **98**, 107204
[11] Li Y, Pan B, Li S, Tong W, Ling L, Yang Z, Wang J, Chen Z, Wu Z and Zhang Q 2014 *New J. Phys.* **16**, 093011

IOP Conf. Series: Journal of Physics: Conf. Series **828** (2017) 012014 doi:10.1088/1742-6596/828/1/012014

# Evidence for spin liquid ground state in $SrDy_2O_4$ frustrated magnet probed by $\mu$SR

**N Gauthier[1,*], B Prévost[2], A Amato[3], C Baines[3], V Pomjakushin[4], A D Bianchi[2], R J Cava[5] and M Kenzelmann[1,†]**

[1] Laboratory for Scientific Developments and Novel Materials, Paul Scherrer Institut, Villigen, Switzerland [2] Département de Physique, Université de Montréal, Montréal, Canada
[3] Laboratory for Muon Spin Spectroscopy, Paul Scherrer Institut, Villigen, Switzerland
[4] Laboratory for Neutron Scattering and Imaging, Paul Scherrer Institut, Villigen, Switzerland
[5] Department of Chemistry, Princeton University, Princeton, New Jersey, USA

E-mail: * nicolas.gauthier@psi.ch, † michel.kenzelmann@psi.ch

**Abstract.** Muon spin relaxation ($\mu$SR) measurements were carried out on $SrDy_2O_4$, a frustrated magnet featuring short range magnetic correlations at low temperatures. Zero-field muon spin depolarization measurements demonstrate that fast magnetic fluctuations are present from $T = 300$ K down to 20 mK. The coexistence of short range magnetic correlations and fluctuations at $T = 20$ mK indicates that $SrDy_2O_4$ features a spin liquid ground state. Large longitudinal fields affect weakly the muon spin depolarization, also suggesting the presence of fast fluctuations. For a longitudinal field of $\mu_0 H = 2$ T, a non-relaxing asymmetry contribution appears below $T = 6$ K, indicating considerable slowing down of the magnetic fluctuations as field-induced magnetically-ordered phases are approached.

## 1. Introduction

In magnetically frustrated systems, competing interactions can inhibit long range order at low temperatures and enforce highly degenerate ground states. This can give rise to novel states of matter such as spin liquids, in which strongly fluctuating magnetic degrees of freedom persist to the lowest temperatures [1]. For example, in the spin ices $Dy_2Ti_2O_7$ and $Ho_2Ti_2O_7$, well-known classical spin liquids, the low temperature dynamics are explained by the presence of magnetic monopoles [2–5]. Other examples include $Tb_2Ti_2O_7$, where the magnetoelastic coupling is believed to play a major role in the spin liquid ground state [6], and $MnSc_2S_4$, where the spins fluctuate collectively between degenerate spiral states [7, 8]. Muon spin relaxation ($\mu$SR) technique is a useful tool to study these types of systems since the experimental signatures of fluctuating moments, static disordered moments and static ordered moments are clearly distinguishable. It is therefore possible to distinguish a spin liquid from a spin glass or a long range ordered system.

The $SrLn_2O_4$ ($Ln$ = Gd, Tb, Dy, Ho, Er, Tm and Yb) compounds form a family of frustrated rare-earth magnets [9, 10]. They crystallize in an orthorhombic structure (space group *Pnam*) with two inequivalent rare-earth ion sites, each forming one-dimensional zig-zag chains along the $c$-axis. In most of these compounds, the crystal electric fields generate different spin anisotropies at the two sites [11, 12] and lead to very anisotropic properties, as seen in magnetic susceptiblity measurements [13]. At very low temperatures, one dimensional correlations are observed in many

IOP Conf. Series: Journal of Physics: Conf. Series **828** (2017) 012014 doi:10.1088/1742-6596/828/1/012014

members of this family [11, 14–16] as a result of dominant magnetic interactions along the chains. The zig-zag chain is equivalent to the one-dimensional chain with nearest neighbour interaction $J_1$ and next-nearest neighbour interaction $J_2$, which for Ising spins is referred to as the axial next-nearest neighbour Ising (ANNNI) model [17]. The model is frustrated for antiferromagnetic $J_2$ and this is the source of magnetic frustration in the $SrLn_2O_4$ compounds. In $SrDy_2O_4$, no evidence for a magnetic phase transition has been reported down to $T = 50$ mK in zero magnetic field despite the emergence of broad magnetic scattering features below $T \approx 4$ K in powder neutron diffraction [9], reflecting the presence of a high degree of frustration. The specific heat features a broad maximum near $T = 1.2$ K, indicative of magnetic correlations, but no sharp peak that would indicate a second order phase transition. At $T = 50$ mK, powder neutron diffraction indicate that the system has quasi-one dimensional (1D) short range correlations [11].

## 2. Experimental details

The $\mu$SR technique measures the time-dependent depolarization of spin-polarized muons that are implanted in a sample. The experimentally probed parameter is the asymmetry $A(t)$ given as:

$$A(t) = \frac{\alpha N_F(t) - N_B(t)}{\alpha N_F(t) + N_B(t)} \tag{1}$$

where $N_F(t)$ and $N_B(t)$ are the number of detected positrons, arising from the weak-decay of the muon, in the forward (F) and backward (B) detectors respectively and $\alpha$ is a parameter taking into account the detector efficiency and geometry. The asymmetry $A(t)$ is proportional to the muon spin depolarization function $P(t)$ and $A(0) = a_0$ is the total initial asymmetry which is in part dependent on the instrument characteristics and typically $a_0 \approx 0.25$.

$\mu$SR spectra have been measured on the General Purpose Surface-Muon Instrument (GPS) and the Low Temperature Facility Instrument (LTF) at the Paul Scherrer Institut on powder samples of $SrDy_2O_4$, prepared as reported in Ref. [9]. The spectra were collected in zero field from $T = 5$ K to 300 K on GPS and from 20 mK to 9 K on LTF. Longitudinal field spectra have been measured on LTF up to $\mu_0 H = 2$ T. The $\alpha$ parameter was determined on GPS from a weak transverse field (TF) spectrum at $T = 300$ K with $\mu_0 H = 0.005$ T, giving $\alpha = 1.269(1)$. For LTF, it was determined from a weak TF spectrum at $T = 20$ mK with $\mu_0 H = 0.010$ T, giving $\alpha = 1.328(3)$. It is noteworthy that our measurements showed no temperature dependence of the TF spectra from $T = 20$ mK up to 9 K on LTF. The obtained $\alpha$ parameter was used in the analysis of the zero-field (ZF) measurements. The value of this parameter changes in longitudinal field (LF) measurements due for instance to the trajectory curvature of the decay positrons in magnetic field. For the results presented here, the field dependence of $\alpha$ was evaluated by fitting it for the spectra at $T = 9$ K in the paramagnetic regime from $\mu_0 H = 0$ to 2 T. The resulting values were used as fixed parameters for the other measured temperatures.

## 3. Experimental results

At $T = 300$ K in zero field, we observe that the initial asymmetry is close to 0.25 and decays exponentially with time (Fig. 1a). The relaxation rate increases with decreasing temperature. Below $T = 60$ K the muon spin depolarization is so fast that most of the asymmetry relaxation occurs in the electronics dead time. This behaviour is observed down to $T = 5$ K, the lowest reached temperature on GPS. A stretched exponential is the most appropriate function to describe the data over the full temperature range and the experimental spectra were fitted by:

$$A(t) = a_s e^{-(\lambda t)^\beta} + a_b \tag{2}$$

where $\lambda$ is the relaxation rate, $\beta$ is an exponent stretching the exponential function, $a_s$ is the sample initial asymmetry and $a_b$ is a constant asymmetry arising from muons missing the sample,

IOP Conf. Series: Journal of Physics: Conf. Series **828** (2017) 012014 doi:10.1088/1742-6596/828/1/012014

with $a_s + a_b = a_0$. The spectra from $T = 60$ K to 300 K were fitted simultaneously and the asymmetry values $a_s$ and $a_b$ were taken as temperature-independent parameters. The fits show that the relaxation rate $\lambda$ increases with decreasing temperature (Fig. 1b) while the exponent $\beta$ is almost one at $T = 300$ K and decreases with decreasing temperature (Fig. 1c).

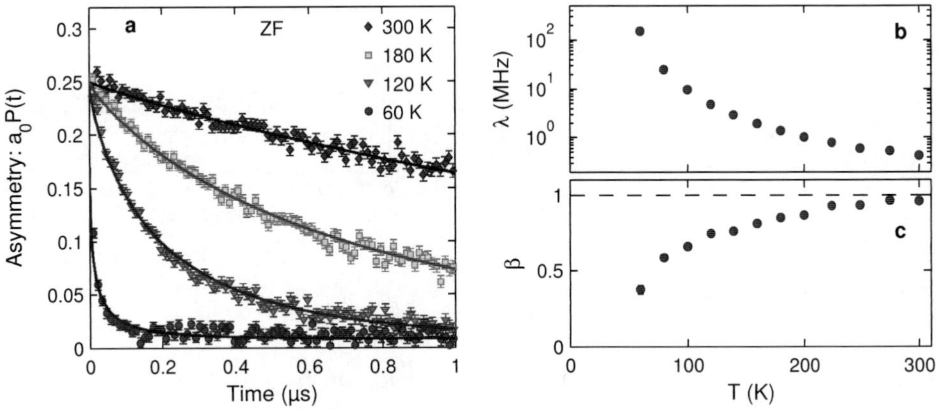


**Figure 1.** (a) $\mu$SR spectra of $SrDy_2O_4$ in zero field at various temperatures on GPS and the temperature dependence of the fitted parameters from Eq. 2: (b) the relaxation rate $\lambda$ and (c) the exponent $\beta$.

For $\beta = 1$, the relaxation corresponds to a simple exponential where the relaxation rate can be defined in the fast fluctuations limit as $\lambda = 2\Delta^2/\nu$ [18]. The parameter $\Delta$ is proportional to $\sqrt{\langle B_{\rm loc}^2 \rangle}$, where $B_{\rm loc}$ is the dynamic local internal field at the muon implantation site, and it thus is a measure of the field distribution width. The parameter $\nu$ is the fluctuation rate of $B_{\rm loc}$. In systems with large magnetic moments, $\Delta$ can be expected to be large and responsible for the fast depolarization of the muon spin. In the paramagnetic regime, $\Delta$ should be weakly temperature dependent while $\nu$ should reduce at lower temperature, as a decreasing number of crystal-field levels are accessible by thermally-induced transitions. This implies an increasing $\lambda$ with decreasing temperature, in agreement with our observation.

A value of $\beta$ different from unity can be interpreted has the presence of a distribution of relaxation rates $\lambda$ [19]. At $T = 300$ K the exponent $\beta$ is close to one and this system is well described by a single relaxation rate. As the temperature decreases, $\beta$ decreases, suggesting a broadening of the relaxation rate distribution, from multiple fluctuation rates $\nu$ and/or multiple field distribution widths $\Delta$. The presence of two inequivalent magnetic sites with different single ion anisotropy could explain such an effect.

$\mu$SR spectra were also measured at LTF from $T = 9$ K down to 20 mK and no significant temperature dependence was observed (Fig. 2a). The temperature range overlaps with the GPS measurements and suggest that the muon spin polarization on the LTF spectra is too fast to be reliably measured. This indicates that fluctuations remain down to 20 mK, in agreement with the absence of long range order in this compound [11].

Longitudinal field measurements give more insights into the low temperature spin dynamics. Typically for static internal fields, the application of a sufficiently strong longitudinal field will dominate the muon spin depolarization and effectively decouple it from the sample internal fields. For dynamic internal fields, the application of a longitudinal field usually has a smaller effect. In $SrDy_2O_4$, the muon spin depolarization is weakly affected by a longitudinal field of $\mu_0 H = 1$ T while a non-relaxing asymmetry contribution attributed to static moments emerges at $\mu_0 H = 2$ T (Fig. 2b). This is strong evidence of the presence of spin dynamics for $\mu_0 H \leq 1$ T.

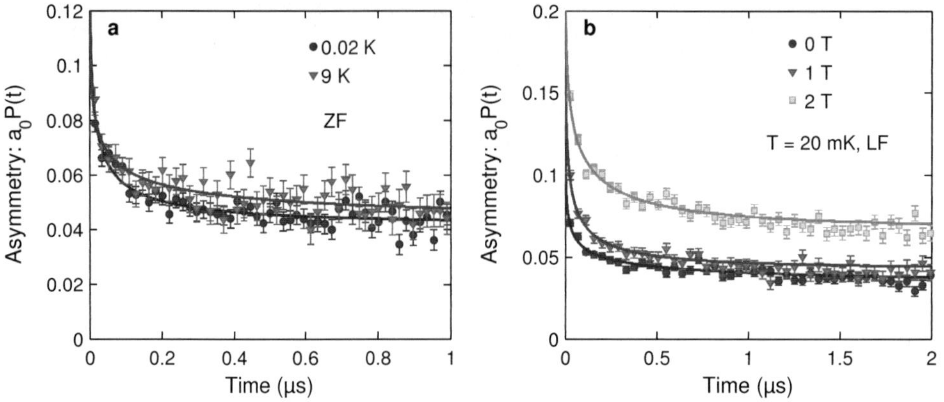


**Figure 2.** $\mu$SR spectra of $SrDy_2O_4$ (a) in zero field at $T = 20$ mK and 9 K on LTF and (b) in longitudinal fields at $T = 20$ mK on LTF.

Longitudinal-field measurements at $T = 9$ K for fields up to $\mu_0 H = 2$ T, shown in Fig. 3a, show the presence of spin dynamics, as expected in the paramagnetic regime. The initial asymmetry increases with increasing field, indicating that the relaxation rate is reduced with field. The spectra were fitted with Eq. 2 by fixing $a_0 = 0.25$, the expected total initial asymmetry, and $a_b = 0.0418$, as determined from the amplitude of the oscillations in the weak TF spectrum at $T = 0.02$ K in zero field. The fitted parameters $\lambda$ and $\beta$ evolve smoothly with field (Fig. 3b,c). The increasing value of the exponent $\beta$ suggests a narrowing of the relaxation rate distribution. The decreasing value of the relaxation parameter $\lambda$ indicates a gradual decoupling of the muon spin polarization from the sample, while the polarization is increasingly affected by the external field. The effect observed here is relatively weak in comparison to the strength of the applied field, indicating that the magnetism at $T = 9$ K is dynamic up to $\mu_0 H = 2$ T.

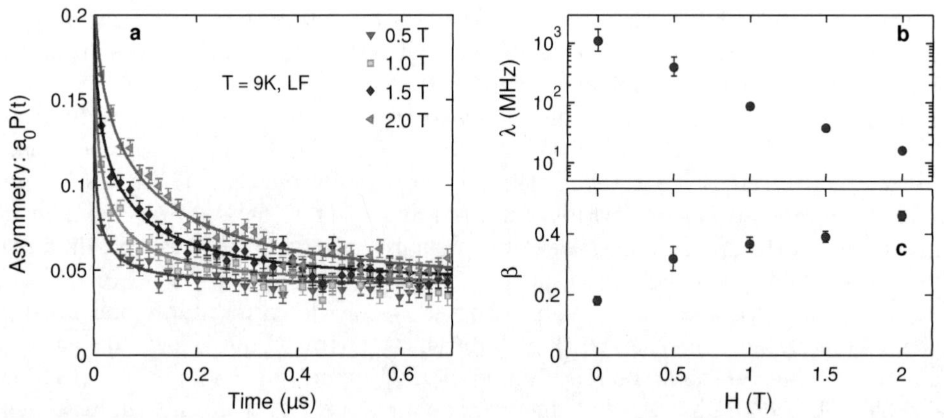


**Figure 3.** (a) $\mu$SR spectra of $SrDy_2O_4$ at $T = 9$ K in longitudinal fields on LTF and the magnetic field dependence of the fitted parameters from Eq. 2: (b) the relaxation rate $\lambda$ and (c) the exponent $\beta$.

A temperature dependence of LF spectra at $\mu_0 H = 2$ T shows the emergence of the non-relaxing asymmetry contribution below $T = 6$ K (Fig. 4). For static fields the muon polarization at long times should lead in a powder sample to a constant $\frac{1}{3}$ of the initial asymmetry. The observed non-relaxing asymmetry is therefore attributed to a sample fraction where the moments appear static to the muons, i.e. the fluctuation rate is too slow to be felt by the muons which have a lifetime of 2.2 $\mu$s. Therefore, the Eq. 2 was modified to take into account this new term:

$$A(t) = a_{\rm s}e^{-(\lambda t)^{\beta}} + a_{\rm static} + a_{\rm b} \tag{3}$$

where $a_{\rm static}$ is a non-relaxing constant accounting for the static fraction of the sample, and the total asymmetry is fixed such that $a_{\rm s} + a_{\rm static} + a_{\rm b} = a_0$. The value $a_{\rm b}$ was taken from the zero field spectrum and kept fixed for all temperatures. The temperature dependence of the stretched exponential parameters are shown in Fig. 4b,c. The exponent $\beta$ is fairly constant but the relaxation rate $\lambda$ increases below $T = 6$ K. Under that same temperature, the constant $a_{\rm static}$ becomes non-zero and increases with decreasing temperature, reaching a value of about one tenth of the initial polarization (Fig. 4d).

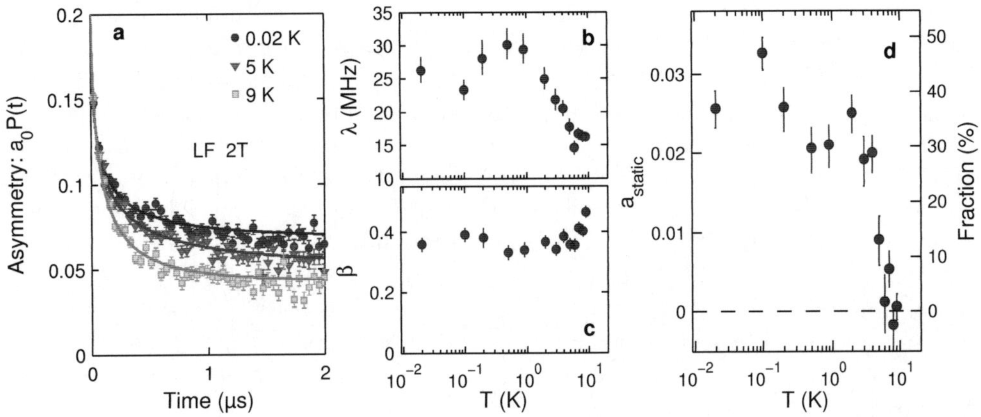


**Figure 4.** (a) $\mu$SR spectra of $SrDy_2O_4$ at various temperatures in a longitudinal field of $\mu_0 H = 2$ T on LTF and the temperature dependence of the fitted parameters from Eq. 3: (b) the relaxation rate $\lambda$, (c) the exponent $\beta$ and (d) the time-independent constant $a_{\rm static}$ (left scale). The corresponding sample fraction which has static moments is also represented on the right scale.

## 4. Discussion and summary

The high temperature $\mu$SR spectra of $SrDy_2O_4$ share many similarities with the spin ice $Dy_2Ti_2O_7$. In both case, a stretched exponential describe the data well and the fitted parameters have a similar tendency with decreasing temperature: (i) an increase of the relaxation rate $\lambda$ and (ii) a reduction of $\beta$ with a maximum of unity at room temperature [20]. Below $T \approx 60$ K, the muon spin depolarization is so fast that it can not be measured in both experiments, probably as a consequence of a broad internal field distribution (large $\Delta$) due to the large magnetic moment of $Dy^{3+}$ of the order of 8-10 $\mu_B$ [11].

At low temperatures, $SrDy_2O_4$ and $Dy_2Ti_2O_7$ show a very different muon spin relaxation. While all the muon spin is mostly depolarized below $T \approx 60$ K, $Dy_2Ti_2O_7$ features a recovery of about one third of the spin polarization below $T \approx 40$ K. This was explained by the dominance of slow magnetic fluctuations that appear static in the time window of the muon decay [20]. Firstly, this leads to a large field distribution at the muon sites and the associated fast relaxation of two thirds of the muon polarization. Secondly, this also leads to a pinning of the muon spin parallel to the magnetic field at the muon site, recovering one third of the muon polarization. This recovered muon polarization decays on a much longer time scale with a relaxation rate directly proportional to magnetic fluctuations frequency.

In contrast, the muon spin polarization in $SrDy_2O_4$ features a fast relaxation in excess of 150 MHz at all temperatures below $T = 60$ K, showing the presence of fast magnetic fluctuations. The origin of these magnetic fluctuations may be thermal excitations to the low-lying crystal-field levels at somewhat intermediate temperatures, and short-range magnetic correlations at lower temperatures, also observed by neutron powder diffraction [11]. The longitudinal field measurements also confirms that fluctuations are present for $\mu_0 H \leq 1$ T at $T = 20$ mK, where thermal fluctuations are not expected to play any role.

For a longitudinal field of $\mu_0 H = 2$ T, the non-relaxing polarization emerging below $T = 6$ K reaches about one tenth of the initial polarization at $T = 20$ mK, corresponding to a sample fraction of $\sim 35\%$ with static moments. These static moments could be ordered or disordered but field-induced phases evidenced by magnetization plateaus [13] suggest the former. The fact that only part of the sample appears static is probably the result of the highly anisotropic properties of $SrDy_2O_4$ [13].

In summary, we present the results of a $\mu$SR study on $SrDy_2O_4$ indicating the presence of fluctuations down to the $T = 20$ mK and the emergence of a static fraction in longitudinal field of $\mu_0 H = 2$ T. Due to the large moments of $Dy^{3+}$, broad field distributions are expected at the muon sites leading to a very fast muon spin depolarization. In the present study, this depolarization becomes too fast to be measured accurately below $T = 60$ K in zero field. The absence of a slowly-relaxing muon spin component on the order of one-third of the initial polarization at low temperatures is evidence that fast fluctuations remain the dominant mechanism for the muon spin relaxation in $SrDy_2O_4$, suggesting a spin liquid ground state.

## Acknowledgments

The authors acknowledge useful discussions with Hubertus Luetkens and Daniel G. Mazzone. This research received support from the Natural Sciences and Engineering Research Council of Canada (Canada).

## References

[1] Balents L 2010 *Nature* **464** 199–208
[2] Castelnovo C, Moessner R and Sondhi S L 2008 *Nature* **451** 42–5
[3] Jaubert L D C and Holdsworth P C W 2009 *Nat. Phys.* **5** 258–61
[4] Fennell T, Deen P P, Wildes A R, Schmalzl K, Prabhakaran D, Boothroyd A T, Aldus R J, McMorrow D F and Bramwell S T 2009 *Science* **326** 415–7
[5] Morris D J P *et al* 2009 *Science* **326** 411
[6] Fennell T *et al* 2014 *Phys. Rev. Lett.* **112** 017203
[7] Bergman D, Alicea J, Gull E, Trebst S and Balents L 2007 *Nat. Phys.* **3** 487–91
[8] Gao S *et al* 2016 *Preprint* arXiv:1605.04199
[9] Karunadasa H, Huang Q, Ueland B G, Lynn J W, Schiffer P, Regan K A and Cava R J 2005 *Phys. Rev.* B **71** 144414
[10] Li H F *et al* 2014 *Front. Phys.* **2** 1–10
[11] Fennell A *et al* 2014 *Phys. Rev.* B **89** 224511
[12] Malkin B Z *et al* 2015 *Phys. Rev.* B **92** 094415
[13] Hayes T J, Young O, Balakrishnan G and Petrenko O A 2012 *J. Phys. Soc. Japan* **81** 024708
[14] Hayes T J, Balakrishnan G, Deen P P, Manuel P, Chapon L C and Petrenko O A 2011 *Phys. Rev.* B **84** 174435
[15] Young O, Wildes A R, Manuel P, Ouladdiaf B, Khalyavin D D, Balakrishnan G and Petrenko O A 2013 *Phys. Rev.* B **88** 024411
[16] Wen J J, Tian W, Garlea V O, Koohpayeh S M, McQueen T M, Li H F, Yan J Q, Rodriguez-Rivera J A, Vaknin D and Broholm C L 2015 *Phys. Rev.* B **91** 054424
[17] Selke W 1988 *Phys. Rep.* **170** 213–64
[18] Dalmas de Réotier P and Yaouanc A 1997 *J. Phys. Condens. Matter* **9** 9113
[19] Campbell I A, Amato A, Gygax F N, Herlach D, Schenck A, Cywinski R and Kilcoyne S H 1994 *Phys. Rev. Lett.* **72** 1291–4
[20] Lago J, Blundell S J and Baines C 2007 *J. Phys. Condens. Matter* **19** 326210

# NMR study on the quasi-one dimensional antiferromagnet $BaCo_2Si_2O_7$

**R. Uesugi, M. Akaki[1], M. Hagiwara[1], H. Kuwahara and T. Goto**

Physics Division, Sophia University, 7-1 Kioicho Chiyodaku Tokyo 102-8551 Japan
[1]Center for Advanced High Field Science, Graduate School of Science, Osaka University, Toyonaka, Osaka 560-8531, Japan

**Abstract**. NMR study has been performed on the quasi-one dimensional antiferromagnet $BaCo_2Si_2O_7$, in which three inequivalent $CoO_4$ tetrahedra, slightly tilted from one another, are linked at vertices to form a chain along the easy-axis *c*. In its magnetically ordered state at 3.6 K, $^{59}$Co-NMR spectra were measured in the field region up to 12 T. From the observed three distinct sets of quadrupolar-split peak groups, three different hyperfine fields at Co sites were determined. They are slightly tilted from *c*-axis and are all different both in magnitude and direction, which is in agreement with the non-collinear magnetic structure.

## 1. Introduction

The relation between magnetic and crystal structure is a fundamental problem in magnetism. The chiral spin structure found recently in the B20 lattice such as MnSi demonstrates its importance[1,2], but on the contrary the simple collinear two-sublattice spin structure was found in $CuSiO_3$ with the spiral crystal structure[3,4]. The key parameter to determine the effect of the lattice is the Dzyaloshinsky-Moriya (DM) interaction, which tends to cant antiferromagnetic spins to form a non-collinear spin structure, which attracts much interest recently, because it is expected to cause a charge anomaly, and hence the multiferroics[5-7]. The title compound $BaCo_2Si_2O_7$ is a quasi-one dimensional antiferromagnet with $T_N$ = 21 K, consisting of vertex-sharing three inequivalent $CoO_4$ tetrahedra, which are tilted from one another (Fig. 1) and are connected to form a chain along the *c*-axis[8]. So far, an intensive study by macroscopic measurements on the compound has revealed that it shows an easy-axis type anisotropy along the *c*-axis including an enhancement in the paramagnetic susceptibility along the *c*-axis and a weak ferromagnetic magnetization in the ordered state below $T_N$ at nearly zero field [**8**,9]. Though the latter seems to be quite consistent with the anisotropy, the $d^7$ configuration of Co atom in the tetrahedral environment leads usually to the isotropic spin state rather than the Ising-type one. The spin structure of the system is not reported until now. So, in order to investigate microscopically the spin state in this compound, we have performed NMR study in its ordered state. From the analysis of $^{59}$Co-spectra in the wide range of applied field 0-12 T, successful extraction of the hyperfine field at each inequivalent Co site will be demonstrated.

## 2. Experimental

The single crystalline sample was prepared by floating zone method [9]. The sample quality, as well as the macroscopic behavior was investigated by X-ray diffraction, specific heat, magnetic susceptibility[9]. The crystal structure of the compound belongs to the monoclinic space group of *C*2/*c* [8]. For the angle $\beta = 90.299°$ is close to 90°, the shape of unit cell can be considered as a cuboid.

For $^{59}$Co-NMR measurements, spectra were obtained by recording the spin-echo amplitude against the magnetic field ($H \parallel c$), which was slowly ramped between 0 and 12 T. The resonance frequency was kept constant while ramping the field, and was changed between 44 and 110 MHz with a step of 1 MHz[10,11]. As the $^{59}$Co nucleus ($I = 7/2$) possesses the electric quadrupole moment, the quadupolar splitting sensitively reflects the local environment, specifically the electric field gradient (EFG) tensor[12]. For the help for later discussion, we state here in detail the local structure around the Co atom. There are three crystallographically-inequivalent Co sites 1, 2 and 3. The shape of $CuO_4$ tetrahedra for each site is slightly different from one another, and also, each tetrahedron is slightly tilted from one another. As shown in Fig. 1, the site 3 includes the two tetrahedra, which are connected with the inversion symmetry, denoted as 3 and 3'. The line-ups of these Co sites along the two inequivalent chains $\alpha$ and $\beta$ in the unit cell are 3-1-3'-2-3 and 3'-2-3-1-3', respectively. The EFG tensor for each Co site is calculated by the point charge model (PCM) approximation[13] and shown in Table 1. Reflecting the tilting of each tetrahedron, all the Co sites have different eigenvalues and principal axes. These tensors determines the quadupolar split in each spectrum as described later. Note that each two corresponding Co sites in the chain $\alpha$ and $\beta$ have the identical EFG tensor. This means that the NMR signal from the corresponding two chains must be degenerated as long as the hyperfine field is the same.

## 3. Results and Discussion

Figure 2 shows field-swept NMR spectra measured with different resonance frequencies $\omega_0$ = 43.83 – 113.83 MHz. One can immediately find major three peak groups A, B and C, which are all quadupolar split. The mass center for each group was extracted and plotted in Fig. 3, which shows the relation between $\omega_0$ and the field, where the signal is observed. There are the two branches of $\omega_0$, the one, which increases monotonically with increasing $H_0$, and the other, which decreases with increasing $H_0$ and takes a minimum and increases again, forming a parabola-like shape. There is no jump in data points up to 12 T, indicating that there is no spin-flip transitions in this field region. Carefully looking parabolas, one finds that it is slightly asymmetric, that is, the gradient at high field side is a little bit larger than that in low field.

We analyze this field-dependence of $\omega_0$ with a quite simple idea in the following. In general, NMR signal is observed when the total field $|\vec{H}_\mu + \vec{H}_0|$ matches $\omega_0/\gamma$, where $\gamma$ =10.03 (MHz/T) is the nuclear gyromagnetic ratio of $^{59}$Co nucleus, $\vec{H}_0 = (0,0,H_0)$, the applied field, and $\vec{H}_\mu$, the hyperfine field at Co site produced by $\vec{\mu}$, the ordered moment of Co 3$d$-spins. If we assume the collinear and two-sub-lattice spin structure along the $c$-axis, that is, $\vec{H}_\mu = (0,0,\pm H_\mu)$, then $\omega_0$ should trace the oblique rectangle shape shown by dashed lines in Fig. 3, which apparently does not accord with the observed $\omega_0$. In order to refine the model to reproduce the data, we take into account the two simple assumptions. One is that $\vec{H}_\mu$ may be slightly tilted from the $c$-axis as

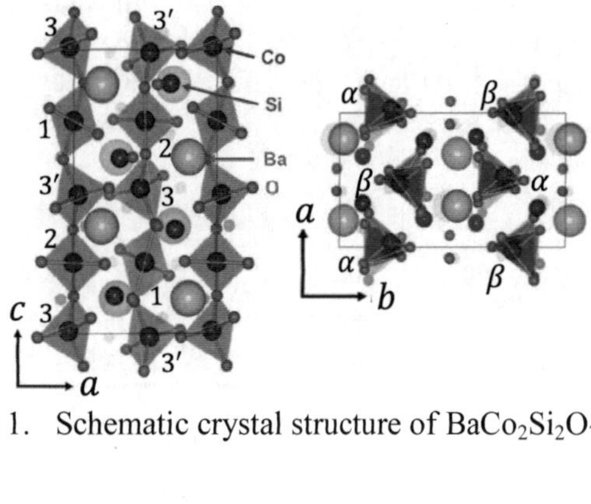


Fig 1. Schematic crystal structure of $BaCo_2Si_2O_7$.

| Co | The largest eigenvalue of EFG tensor and its principal axis |
|---|---|
| 1 | 0.0463 (+0.8815, 0, –0.4721) |
| 2 | 0.2132 (+0.9946, 0, –0.1039) |
| 3 | 0.1600 (+0.7682, ±0.457, +0.4483) |

Table 1. The largest eigenvalue (in the unit of esu/Å$^2$) and its principal axis of EFG, calculated by PCM for each Co site.

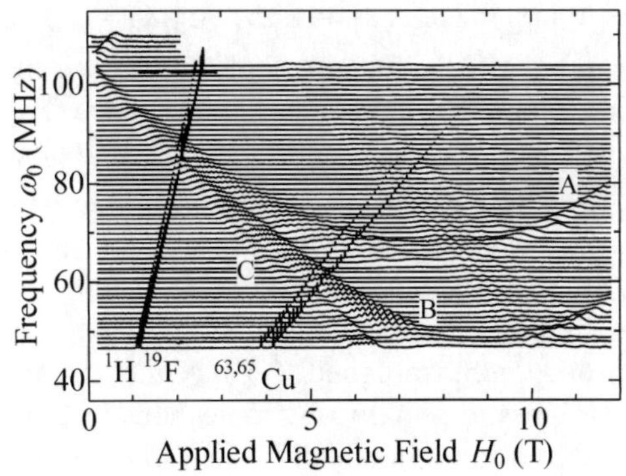


Fig 2. Field-swept spectra of $^{59}$Co-NMR measured with various frequencies $\omega_0$. Each spectrum is drawn shifted in proportion to $\omega_0$. The small sharp peaks are spurious signal of $^1$H, $^{19}$F and $^{63/65}$Cu coming from the probe.

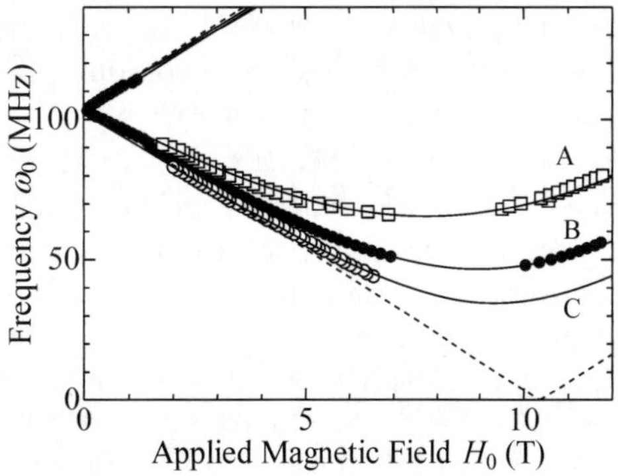


Fig. 3 Signal positions of three Co sites for various NMR frequencies $\omega_0$. The dashed line shows the collinear model, and the solid curves, non-collinear one. The fitted parameters are shown in Table 2.

$\vec{H}_\mu = (H_\mu \sin\theta \cos\phi\,, H_\mu \sin\theta \sin\phi\,, \pm H_\mu \cos\theta)$, where $\phi$ and $\theta$ are polar coordinates, and the plus and minus signs correspond to moments in two-sublattice, that is, the positive sign is taken for the one with the hyperfine field oriented along the applied field, and the negative sign, in the opposite direction. Another assumption is that with increasing $H_0$, $\theta$ may change from its zero-field value $\theta_0$ as $\theta = \theta_0 + aH_0$ $(\theta_0 + aH_0)$, where $a$ is a small proportional constant, for the sublattice with the hyperfine field oriented along the applied field (the opposite direction), respectivfely. This effect reflects the fact that with increasing field along the $c$-axis, the ordered moments change their direction slightly.

By adjusting those three constants $\theta_0$, $a$ and $H_\mu$, observed field dependence of $\omega_0$'s for the three Co sites is well reproduced as shown by solid curves in Fig. 3. Especially, note that the field dependence for the signal from each sublattice was successfully reproduced; the data points below (above) 100 MHz corresponds to the moments with the $c$-component of the hyperfine field, $-H_\mu \cos\theta$ $(+H_\mu \cos\theta)$. Obtained parameters are shown in Table 2. Considering the relative amplitude of signal, one can safely assign the group B to the Co site 3, which has twice a fraction of the other two. Thus, the group A and C are assigned either 1 or 2.

The assumption of tilted hyperfine field model can be justified by following three observations. First, the value of $H_\mu \simeq 10$ T for all three Co sites is reasonable for divalent transition metals. Next, the model well explains the fact that signal disappears at the parabola's bottom, where the total field becomes parallel with $h_1$, the oscillating field for the NMR measurement. The third point is that the quadupolar splitting width is found to be field dependent. For the quadupolar splitting width is proportional to $1 - 3\cos^2\Theta$, where $\Theta$ is the angle between the total field $\vec{H}_\mu - \vec{H}_0$ and the principal axis of EFG, shown in Table. 1, one can see that the width should change with the applied-field only when $\vec{H}_\mu$ is tilted from $\vec{H}_0$, hence from the $c$-axis. Thus, we can conclude here that the hyperfine field at each Co site is non-collinear, which suggests the non-collinear spin structure.

| Co | $H_\mu$(T) | $\theta_0$(deg) | $a$(deg/T) |
|---|---|---|---|
| 1 or 2 | 10.3(1) | 36(2) | 0.45(3) |
| 2 or 1 | 10.2(1) | 15(2) | 0.40(3) |
| 3 | 10.4(1) | 22(2) | 0.50(3) |

Table 2. The size, direction and coefficient for field dependence of the hyperfine field for each Co site.

Next, we proceed and discuss a possible arrangement of ordered moments. If one accepts the criterion to minimize the antiferromagnetic exchange energy, the spin structure in each chain should be like, 3(+)-1(−)-3'(+)-2(−)-3(+) or 3(−)-1(+)-3'(−)-2(+)-3(−), where signs denote that of the $c$-component in hyperfine field, and hence the sublattice. These two possible patterns may form a domain structure, which can easily be inverted by applying a low field. Within each domain, there is an imbalance in the net $c$-component of $\vec{H}_{\mu}$; this may explain the finite weakferromagnetic moment along the $c$-axis. However, one should note here that the ordered moment $\vec{\mu}$ is not necessarily parallel with the hyperfine field; the two are connected with the hyperfine coupling tensor. The tensor may have an appreciable anisotropy reflecting the shape of $3d$ orbitals. In order to determine the moment direction $\vec{\mu}$ completely, the knowledge of hyperfine coupling tensor is indispensable, and for its determination by so-called $K$-$\chi$ plot method, the NMR measurement in the paramagnetic state is now in progress. Finally, we note here that the azimuth $\phi$'s, which are remained to be undetermined, is expected to be obtained by investigating the applied-field dependence of quadupolar splitting with a higher precision, which is also in progress.

**Summary**

$^{59}$Co-NMR study has been performed on the quasi one dimensional antiferromagnet $BaCo_2Si_2O_7$ in its ordered state. The hyperfine field at each four inequivalent Co sites was determined to be approximately 10 T, and tilted appreciably from the $c$-axis, which is in agreement with the non-collinear-type spin structure.

**Acknowledgements**

This work was supported by JSPS KAKENHI Grant Number 26108713, 15K05148 and 15K05145.

**References**

[1] H. Yasuoka *et al.*, J. Phys. Soc. Jpn 85, 073701 (2016)
[2] S. Mühlbauer *et al.*, Science **323,** 915 (2009)
[3] A. Podlesnyak *et al.*, Phys. Rev. B93, 064426 (2016)
[4] K. Matsui *et al.* JPS Conf. Proc. 3, 014011 (2014)
[5] Y. Tokura *et al.* Rep. Prog. Phys. **77**, 076501 (2014)
[6] H. Katsura *et al.*, Phys. Rev. Lett. **95**, 057205 (2005)
[7] L. N. Bulaevskii *et al.,* Phys. Rev. **B78**, 024402 (2008)
[8] R. D. Adams *et al.*, Inorg. Chem. 35 (1996) 3492.
[9] M. Akaki *et al.*, Physica B 403 (2008) 1505.
[10] K. Kanada *et al.* J. Phys. Soc. Jpn. **76**, 064706 (2007)
[11] A. Oosawa *et al.* J. Phys. Soc. Jpn. **78**, 053702 (2009)
[12] H. Inoue *et al.* Phys. Rev. **B79**, 174418 (2009)
[13] T. Kawase *et al.* Phys. Procedia 75, 641 (2015)

IOP Conf. Series: Journal of Physics: Conf. Series **828** (2017) 012016 doi:10.1088/1742-6596/828/1/012016

# NMR study on the competing spin chain $Rb_2Cu_2Mo_3O_{12}$

**A. Yagi, K. Matsui, T. Goto, and M. Hase[1] and T. Sasaki[2]**

Physics Division, Sophia University, Chiyodaku Tokyo 102-8554 Japan
[1]National Institute for Materials Science (NIMS), Tsukuba Ibaraki 305-0047 Japan
[2]Institute for Material Science(IMR), Tohoku Univ., Sendai 980-8577 Japan

E-mail: gotoo-t@sophia.ac.jp

**Abstract.** Quasi-one-dimensional $S = 1/2$ Heisenberg system $Rb_2Cu_2Mo_3O_{12}$ described by the $J_1$ - $J_2$ model with the nearest-neighbor ferromagnetic and next-nearest-neighbor antiferromagnetic exchange interactions have been investigated by $^{87/85}$Rb-NMR under extended field range up to 18 T. From the thermal-activation-type temperature dependence of $1/T_1$, opening of the spin excitation gap was confirmed under the low field below 2 T. With increasing the field, the system becomes gapless, and the gap opens again in the high field saturation region, where the field dependence of the gap size showed the coherent excitation of the neighbouring two spins.

## 1. Introduction

One dimensional and quasi-one dimensional spin systems have brought and are still giving rich physics, in spite of their simple Hamiltonians. Among them, the $J_1$ - $J_2$ model, which has two dominant exchange interactions of the nearest-neighboring ferromagnetic $J_1$ and the second nearest-neighboring antiferromagnetic $J_2$ [1,2] attracts much interest recently, because of theoretical suggestion of spin-nematic state or spin-nematic Tomonaga-Luttinger liquid state, where vector-natured spins behave like quadrupoles [3-5]. So far, many model compounds are proposed, but an experimental evidence of its detailed spin state is unknown.

In this article, we focus on the quasi-one dimensional magnet $Rb_2Cu_2Mo_3O_{12}$, which can be described by the $J_1$ - $J_2$ model with $J_1 = -138$ K and $J_2 = +51$ K[6,7], and investigate its low-temperature spin state by NMR-$T_1$ under the wide field range up to 18 T. We pick up the two viewpoints for study. One is the magnetic ground state of the system. The investigations on macroscopic quantities reported that the system shows no long range magnetic order down to 0.3 K, and rather has a very small energy gap in the spin excitation spectrum, that is, the gap size is $\Delta = 1.6$ K, which is smeared out by the low field of $H_c = 2.3$ T.[8,9] This makes clear contrast with the isomorphic compound $Cs_2Cu_2Mo_3O_{12}$, which also can be described by the model with slightly different exchange parameters. The latter has been confirmed to exhibit a magnetic order below 1.8 K under zero and finite fields.[10] We will investigate microscopically how the spin state of $Rb_2Cu_2Mo_3O_{12}$ changes with magnetic field.

The second viewpoint is of the saturated region at high fields, where the spins are forced to be aligned ferromagnetically. Generally, in that state, the spin system has a finite energy gap, the size of which corresponds to the Zeeman energy of one-spin-flop state. However, a recent theoretical report suggests that the first excited state of those described by $J_1$ - $J_2$ model may be the two-magnon state.

IOP Conf. Series: Journal of Physics: Conf. Series **828** (2017) 012016 doi:10.1088/1742-6596/828/1/012016

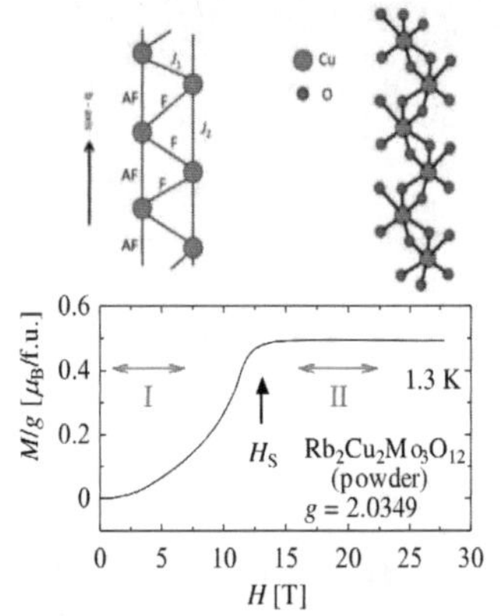


**Figure 1.** Schematic drawing of exchange path for $J_1$-$J_2$ model and the arrangement of $CuO_6$ octahedra (top). $M$-$H$ curve with field regions for NMR measurements, shown by horizontal arrows (bottom).

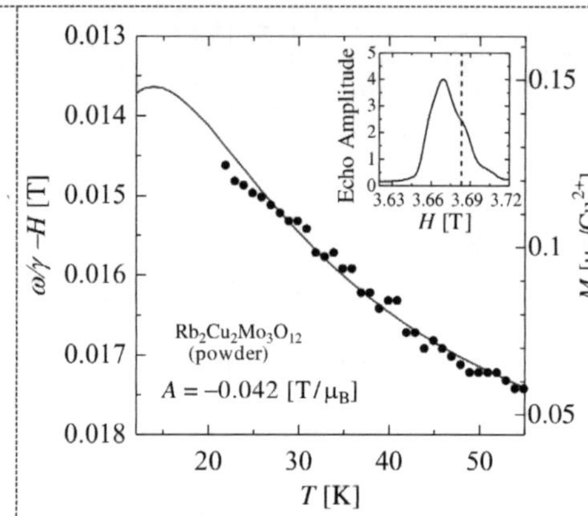


**Figure 2.** The scaled plot of the temperature of the uniform magnetization and the Knight shift. The inset shows a typical spectra corresponding to the central transition line, with the zero shift position denoted by dashed line.

By utilizing the temperature dependence of $T_1$ as a probe, we will demonstrate directly field-dependence of the gap[11-12].

## 2. Experimental

The sample of $Rb_2Cu_2Mo_3O_{12}$ was obtained in powder form by the conventional solid-state reaction method [6,14]. Magnetization measurement shows that the saturation field is approximately $H_s$ =12.7 T[6,9]. For better understanding of NMR results below, we refer the detailed crystal structure here. The space group of the crystal structure is *C*2/*c*, and the edge-sharing $CuO_6$ distorted octahedra form a so-called ribbon chain along *b*-axis (Fig. 1). As for the NMR-nucleus Rb, which are located between interstitial positions between chains, there are three inequivalent sites of 4*d*, 4*e* and 8*f*. The distances between these Rb and their nearest-neighboring Cu sites are within a range between 4.07 and 5.61 Å, indicating that all the Rb sites may have moderate hyperfine coupling with Cu moments.

We have performed NMR measurements mainly in the two regions I and II, the low field region below 5 T and the saturated region 13-18 T, respectively. The two isotopes of $^{87}$Rb ($\gamma = 13.928$, $I = 3/2$) or $^{85}$Rb ($\gamma = 4.1099$, $I = 5/2$) were chosen in measurements at the region I and II, respectively. Spectra were obtained by recording the spin-echo amplitude against magnetic field, which was ramped slowly around the zero-shift position [15,16]. Since the sample was powder, both the *eqQ*-powder pattern of the center line and the peak structure due to the signal from inequivalent sites are smeared out, so that only a single broad peak was observed; a typical spectrum profile is shown in the inset of Fig. 2. The longitudinal spin relaxation rate $1/T_1$ was obtained by the conventional saturation-recovery method with a pulse-train for saturation[17]. The nuclear spin polarization after the saturation was traced until its difference from the thermal equilibrium value reached one percent. The time evolution of the nuclear spin polarization was fitted by the function $0.1e^{-\tau/T_1} + 0.9e^{-6\tau/T_1}$ and $0.029e^{-\tau/T_1} + 0.18e^{-6\tau/T_1} + 0.79e^{-15\tau/T_1}$ for $^{87}$Rb and $^{85}$Rb, respectively.

## 3. Results and Discussion

The temperature dependence of the Knight shift the uniform magnetization are scaled and plotted together in Fig. 2, from which the effective hyperfine coupling constant $A = -420\,\mathrm{Oe}/\mu_B$ was

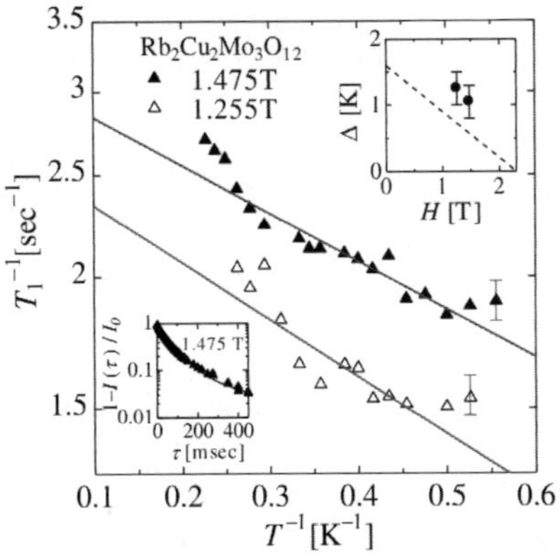


**Figure 3.** The temperature dependence of $1/T_1$ at low field below 1.5 T. The solid lines show the thermal activation law $e^{-\Delta/k_{\mathrm{B}}T}$. The obtained gap $\Delta$ is plotted against the field in the inset, where the dashed line shows the expected behaviour from ref. [9].

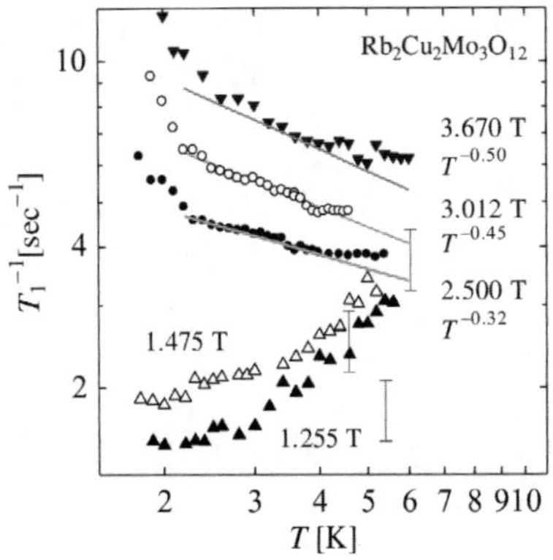


**Figure 4.** The temperature dependence of $1/T_1$ below 4 T (the data in Fig. 3 are also shown for comparison). The solid lines show the power law temperature dependence.

obtained. This value is large enough to study the magnetism in copper moments by NMR. In the following, we describe the results of $1/T_1$ measurements at low (region I) and high (region II) fields. First, we show in Fig. 3, the temperature dependences of $1/T_1$ at lowest fields below 1.5 T, where the opening of gap was reported by the temperature dependence of uniform susceptibility [9]. One can see that $1/T_1$ obeys the thermal activation law, from which a finite gap $\Delta$ was evaluated and plotted against the applied field shown in the inset. The gap size was 1.3 K under the field 1.255 T, and became smaller as the field is increased. This gives a microscopic demonstration on the existence of spin excitation gap at low fields, consistent with Fujimura's report on the uniform susceptibility [9]. Note that the small discrepancy of our gap estimation from the ref. 9 may be due to the fact that our measured temperature range is marginal, that is, comparable to the gap size, and hence the gap size and hence the gap size may be overestimated. Furthermore, at these low fields, the S/N ratio of NMR is quite low and only two field points were obtained (the inset of Fig. 2), which, we admit, are not enough to discuss the field dependence.

Next, we raise the field above 2 T, and plot the temperature dependence of $1/T_1$. The data below 2 T are also shown for comparison. The temperature dependence drastically changes at around 2 T, above which the energy gap seems to be collapsed. This boundary is very close to the reported on the magnetization $H_{\mathrm{c}} = 2.3$ T [9]. We can fit the temperature dependence with the power law $T^{-\alpha}$ within a limited range between 2 and 5 K. The power law index $\alpha$ is field dependent and appreciably increases with increasing field. Though this may be an indicative of TLL state [19-23], one needs more data for the comparison with theories. Furthermore, the information on the anisotropy in the hyperfine tensor is indispensable to extract the Luttinger parameter $K$, which determines the spatial spin correlation, from the temperature dependence of $1/T_1$. This is because the longitudinal and transverse components of spin correlation contribute to $1/T_1$ and cause different temperature dependence as $T^{2K-1}$ and $T^{1/2K-1}$, respectively [19,21]. In order to determine the hyperfine coupling tensor, NMR experiments on axially-aligned powder sample is now in the progress.

Above the higher temperature boundary of 5 K, $1/T_1$ tends to be temperature independent, that is, approaching to the paramagnetic behavior in the high temperature limit. And on the contrary, below the lower boundary, $1/T_1$ starts to diverge toward a possible critical temperature. Though this behavior still does not show any evidence for phase transitions, and there has been no reports on the existence of the field-induced magnetic order in this compound, one can refer at least that the spin gap

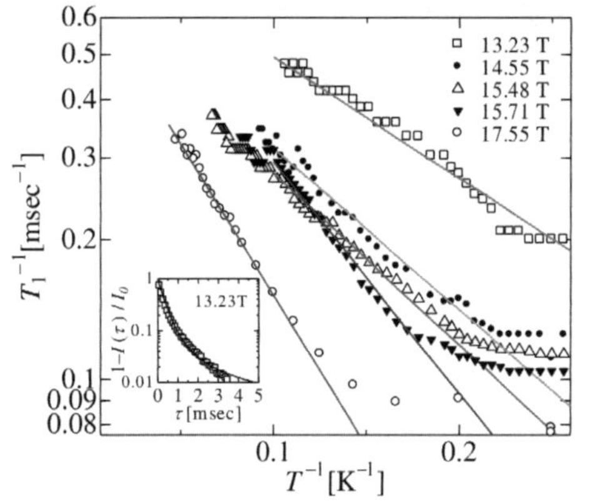


**Figure 5.** The temperature dependence of $1/T_1$ in the saturated region above $H_s = 12.7$ T. The line shows the thermal activation law. The inset shows the typical relaxation curve.

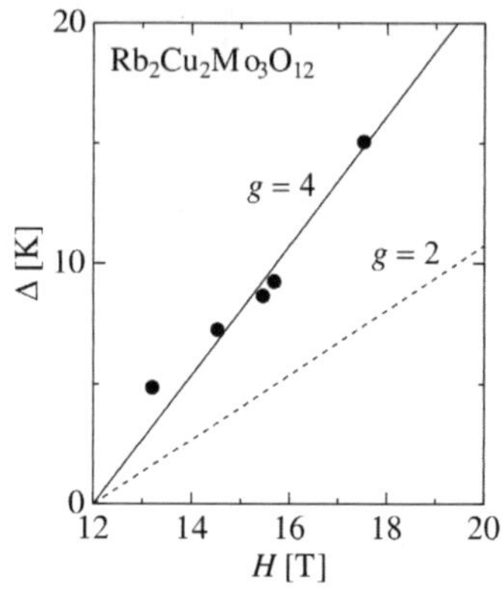


**Figure 6.** Field dependence of energy gap Δ in the saturated region. The solid and dashed curves show the relation $\Delta = g½\mu_B(H - H_s)$, where $H_s = 12.7$ T.

is closed, and the system becomes completely gapless to have the magnetic ground state. The measurement at still lower temperature region is under progress.

Next, we proceed to the high field of saturated region (region II), where $1/T_1$ obeys again the thermal-activation type law as shown in Fig. 5, indicating the existence of spin excitation gap Δ. It's field dependence is shown in Fig. 6, where one can see that the gap increases with the field as $\Delta = g½\mu_B(H - H_s)$, and that $g$ takes an anomalous value of 4. This clearly demonstrates that the two spins flip coherently from the ground state of forced ferromagnet in the saturated region[13]. Evidently, this coherency is considered to come from the ferromagnetic exchange interaction between the neighboring two spins.

**Summary**
NMR-$1/T_1$ measurements on the $S = ½$ Heisenberg competing spin chain system $Rb_2Cu_2Mo_3O$ were performed in the wide-field range from zero to 18 T. Under the low field of 1.255 T, existence of a small spin excitation gap $\Delta \simeq 1.3$ K was confirmed. With increasing field the gap shrank above 2 T, and the ground state became magnetic. In the high field of saturated region, the system becomes gapfull again, and the gap size obeyed the field dependence of $\Delta = g½\mu_B(H - H_s)$ with $g = 4$, indicating the coherent excitation of two-magnons.

**Acknowledgement**
This work was partly supported by JSPS KAKENHI Grant Number 26108713, 15K05148 and. It was also supported by the Sophia University Special Grant for Academic Research. A part of the experiments was performed at High Field Laboratory for Superconducting Materials, Institute for Materials Research, Tohoku University.

**References**
[1] T. Tonegawa *et al.,* J. Phys. Soc. Jpn. 58 (1989) 2902.
[2] C. Itoi *et al.,* Phys. Rev. B63 (2001) 224423.
[3] T. Hikihara et al.: Phys. Rev. B78 (2008) 144404.
[4] M. Hagiwara et al.: J. Phys. Conf. Ser. 320 (2011) 012049.
[5] Lars Kecke et al.: Phys. Rev. B76, 060407R (2007).
[6] M. Hase *et al.,* Phys. Rev. B70 (2004) 104426.

[7] M. Hase *et al.,* J. Appl. Phys. 97 (2005) 10B303.
[8] A. Fujimura *et al.,* IEEE Transactions on Magnetics 52 (2016) 1100503.
[9] Y. Yasui *et al.*, JPS Conf. Proc. 3, 014014 (2014).
[10] Y. Hoshino *et al.,* JPS Conf. Proc. 3 (2014) 014012.
[11] H. Tsunetsugu et al.: J. Phys. Soc. Jap. 75 (2006) 083701.
[12] M. E. Zhitomirsky et al.: Euro Phys. Lett. 92 (2010) 37001.
[13] N. Büttgen *et al.*: Phys. Rev. B90 (2014) 134401.
[14] S. F. Solodovnikov et al.: J. Struc. Chem. 38 (1997) 765.
[15] H. Inoue *et al.* , Phys. Rev. B79 (2009) 174418.
[16] K. Kanada *et al.*, J. Phys. Soc. Jpn. 76 (2007) 064706.
[17] T. Adachi *et al.*, J. Phys. Soc. Jpn. 76 (2007) 083701.
[18] A. Narath, Phys. Rev. 162 (1967) 320.
[19] M. Sato et al.: Phys. Rev. Lett. 110, 077206 (2013).
[20] T. Sakai, J. Phys. Soc. Jpn. 64 (1995) 251.
[21] M. Sato et al. Phys. Rev. B83 (2011) 064405.
[22] M. Klanšjek et al.: Phys. Rev. Lett. 101 (2008) 137207.
[23] K. Ninios et al.: Phys. Rev. Lett. 108 (2012) 097201.

IOP Conf. Series: Journal of Physics: Conf. Series **828** (2017) 012017 doi:10.1088/1742-6596/828/1/012017

# NMR and μ-SR study on competing Heisenberg chain $Cs_2Cu_2Mo_3O_{12}$

**T. Goto, A. Yagi, K. Matsui, I. Watanabe[1], T. Sasaki[2] and M. Hase[3]**

Physics Division, Sophia University, Chiyodaku Tokyo 102-8554 Japan
[1]RIKEN, 2-1 Hirosawa, Wako, Saitama 351-0198, Japan
[2]Institute for Material Science(IMR), Tohoku Univ., Sendai 980-8577 Japan
[3]National Institute for Materials Science (NIMS), Tsukuba Ibaraki 305-0047 Japan

E-mail: gotoo-t@sophia.ac.jp

**Abstract.** $S$ = 1/2 quasi one dimensional Heisenberg chain $Cs_2Cu_2Mo_3O_{12}$ has two dominant exchange interactions of the nearest-neighbouring ferromagnetic $J_1 = -93$ K and the second nearest-neighbouring antiferromagnetic $J_2 = +33$ K. These competing interactions on the spin chain are expected to bring rich physics, such as frustration or nematic Tomonaga-Luttinger liquid (TL) phase. In order to investigate its ground state at zero and finite fields, we have performed $^{133}$Cs-NMR experiments under the wide range of magnetic field up to 14 T, and μ-SR at zero field. In low-field region, the existence of long range magnetic order was demonstrated below 2 K. In the higher field region between 7 T and the saturation field $H_s = 9.15$ T, no evidence for magnetic order was confirmed at finite temperatures.

## 1. Introduction

The ground state of $S$ = 1/2 Heisenberg spin chain is the Tomonaga-Luttinger liquid (TLL) state, where the spatial spin correlation obeys the power law. This is a critical state, and the system may undergo a phase transition at a finite temperature with aid of a tiny inter-chain interaction. We focus attention on the quasi-one dimensional spin chain $Cs_2Cu_2Mo_3O_{12}$ [1], which is isomorphic to $Rb_2Cu_2Mo_3O_{12}$ [2], and is reported to be described by the $J_1$- $J_2$ model with two dominant exchange interactions of the nearest-neighboring ferromagnetic $J_1 = -93$ K and the second nearest-neighboring antiferromagnetic $J_2 = +33$ K [2]. In this competing chain model, the existence of nematic TLL state is theoretically proposed under high field region near the saturation [4-8]. The ground state of $Cs_2Cu_2Mo_3O_{12}$ under zero or finite field is still not clear until now. The uniform susceptibility approaches a finite value at zero temperature without showing an anomaly[1], suggesting a gapless ground state. Quite recently, preliminary results on NMR[9] and specific heat[12] have reported the existence of magnetic order at low temperatures below 2 K under a finite field of 5 T and zero field respectively. In order elucidate microscopically its ground state in a wide range of magnetic field, we have performed μSR and NMR study in fields up to 14 T. In this article we show the field dependence of magnetic order critical temperature, and also the temperature dependence of NMR longitudinal nuclear spin relaxation rate $1/T_1$ in paramagnetic state, which shows a drastic change with applied magnetic field.

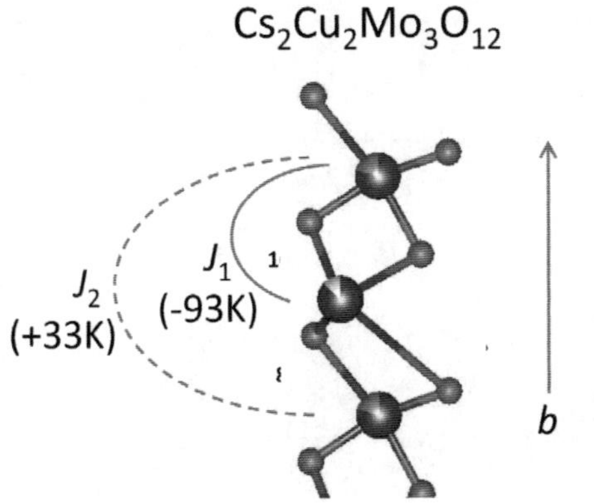


**Figure 1.** Schematic drawing of the ribbon-chain structure consisting of Cu (blue sphere) and Oxygen (red smaller sphere). Exchange paths of $\boldsymbol{J_1}$ and $\boldsymbol{J_2}$ are shown by curves [1].

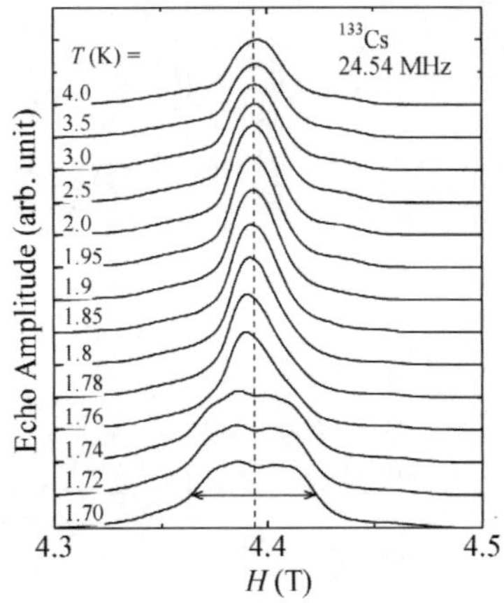


**Figure 2.** Typical profiles of $^{133}$Cs-NMR spectra for powder sample at various temperatures. The horizontal arrow indicates the definition of resonance line width.

## 2. Experimental

The powder sample of $Cs_2Cu_2Mo_3O_{12}$ was obtained by the conventional solid-state reaction method[1]. Its crystal structure belongs to the space group of *C*2/*c* and consists of zigzag chains or so-called ribbon chains [1,3], which run along *b*-axis (Fig. 1) and of intervening ions of Mo and Cs located at interstitial site between chains, maintaining the good one-dimensionality. The NMR-nucleus Cs has the three inequivalent sites of 4*d*, 4*e* and 8*f*, and the distances between Cs and Cu atoms are between 4 and 6 Å, from which a moderate hyperfine coupling is expected.

NMR measurements were performed in the field region between 1.5 and 14 T and in the temperatures down to 0.3 K. Spectra were obtained by recording the spin-echo amplitude against magnetic field, which was ramped slowly around the zero-shift position [13]. For the sample was powder, both the *eqQ*-splitting of $^{133}$Cs ($I = 7/2$) and the signal from inequivalent sites are smeared out, so that only a single broad peak was observed in the paramagnetic temperature region.

The longitudinal spin relaxation rate $1/T_1$ was obtained by the conventional saturation-recovery method with a pulse-train for saturation. We note here that due to an extraordinary long $T_2$, the transverse nuclear spin relaxation time, the distance between each pulse within the train must be set apart by as long as 100-200 $\mu$sec. After the saturation, the nuclear spin polarization was traced until its difference from the thermal equilibrium value reached one percent. Obtained relaxation curves were fitted to the stretched exponential function $1 - e^{-(\tau/T_1)^\beta}$, where $\beta$ is a temperature dependent constant, and its difference from unity reflects microscopic inhomogeneity in the system[9-11]. The effect of quadrupole splitting [15] was observed neither in relaxation nor spectrum experiments, which is considered to be due to the extremely small electric quadrupole moment of $^{133}$Cs nuclei. The constant $\beta$ was nearly temperature independent around 0.5 above 2 K, and was decreased gradually to 0.2 below 2 K [9,14]. The origin of its temperature dependence is not determined at this stage, though it is speculated to be related to microscopic inhomogeneity[10] in this system.

The zero-field (ZF) μSR experiment was performed at RIKEN-RAL muon facility in ISIS, UK. In this experiment, the spin-polarized muon bunch with a momentum of 27 MeV/*c* is injected into the sample and the time-evolution of muon spin polarization was traced until $\tau = 8$ $\mu$sec. The obtained depolarization curves were analyzed with the function $G_{KT}(\tau;\sigma)e^{-\lambda\tau}$, where $G_{KT}$ is Kubo-Toyabe function, $\sigma$, the quasi-static field distribution contributed from nuclear moments and $\lambda$, the

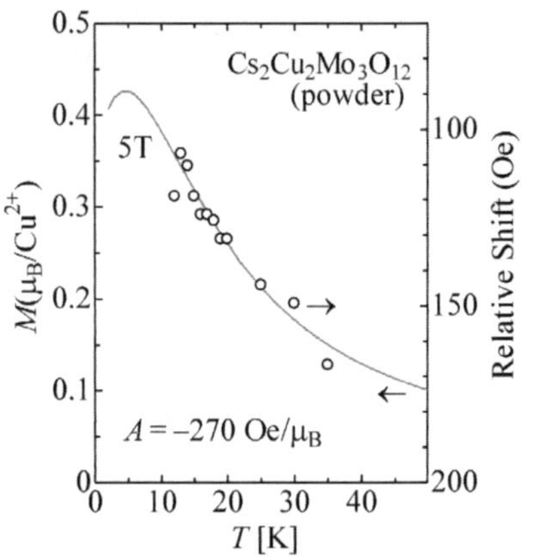


**Figure 3.** Scaled plot of the temperature dependences of NMR shift and uniform magnetization to obtain the effective hyperfine coupling constant $\boldsymbol{A}$.

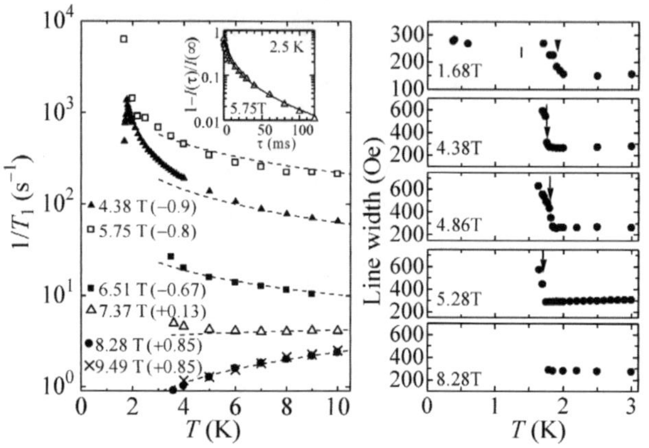


**Figure 4.** (Left) Temperature dependence of the $T_1^{-1}$ in various fields. Dashed curves show the power law $T^{2K-1}$; the values of $2K-1$ are shown with the field values. The inset show typical relaxation curves. (Right) The line width taken at various applied fields. $T_N$ defined as an abrupt increase in the width is shown by vertical arrows.

depolarization rate due to the dynamical spin fluctuation[11,14,16]. In the present experiments, $\sigma \simeq 0.068\ \mu\mathrm{sec}^{-1}$ was found to be temperature independent, and the information of spin state was obtained solely from the temperature dependence of $\lambda$.

## 3. Results and Discussion

Figure 2 shows a typical profile of spectra at various temperatures. With decreasing temperature, the line width slightly decreased below 10 K, reflecting the decrease in uniform susceptibility [1,12], and abruptly increased again below 1.85 K. Figure 3 shows the temperature dependence of the uniform susceptibility and the resonance line shift in the paramagnetic state. By scaling those two quantities, we obtained the effective hyperfine coupling constant $A = -270\ \mathrm{Oe}/\mu_\mathrm{B}$, which is large enough to detect the magnetism in copper site by NMR. Note that this value of $A$ includes the contributions both from the isotropic and anisotropic parts, because the definition of shift is the peak top rather than the mass center of peak.

Next, we show the temperature dependences of $1/T_1$ and the line width in Fig. 4. At the field 4.38 T, a critical divergence in $1/T_1$ [17] and an abrupt increase in the line width took place at the same temperature 1.8 K. This clearly demonstrates the existence of long range magnetic order, which is consistent with our previous report [9]. This finite $T_\mathrm{N}$ in the quasi-one dimensional system is considered to be brought by a weak inter chain interaction. Thus, we can safely assign the increase in the line width at 1.68 and 5.28 T to the magnetic order. At the higher field of 5.75 T, $1/T_1$ showed a clear diverging toward 1.7 K, and we can judge that $T_\mathrm{N}$ must be located in its vicinity, that is, slightly below 1.7 K, though we have observed no increase in the line width down to 1.7 K.

At field 8.28 T, $1/T_1$ monotonically decreased with decreasing temperature without showing any sign of the critical divergence. The line width did not show any abrupt increase down to 1.7 K. These indicate an absence of magnetic order at finite temperatures at this and the higher fields. The data at 7.37 and 6.51 T are, however, critical, because $1/T_1$'s stayed nearly constant, and no broadening was observed (not shown) down to 1.7 K. We can therefore only say that a possible $T_\mathrm{N}$ for these two field points may exist between 1.5 and 0 K.

Zero field μ-SR (ZF) data in Fig. 5 shows that the muon spin depolarization rate $\lambda$ abruptly increased below $T_\mathrm{N} = 1.85$ K. This anomaly coincides with the recent report on the specific heat by

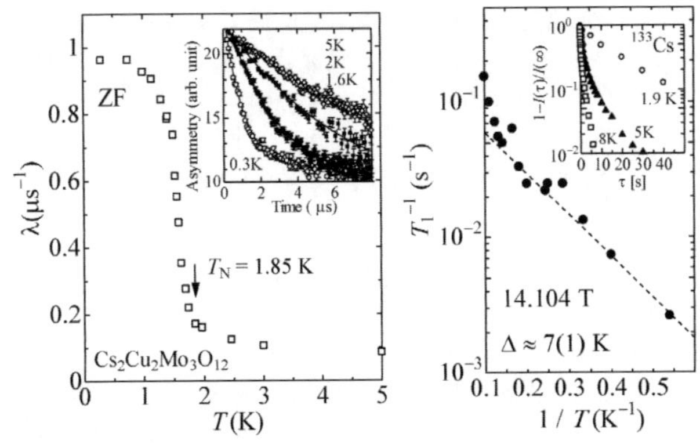


**Figure 5.** (Left) Temperature dependence of μ-SR depolarization rate at zero field. $T_N$ is shown by an arrow. The inset shows typical depolarization curves. (Right) Temperature dependence of $T_1^{-1}$ at high field in the saturated region.

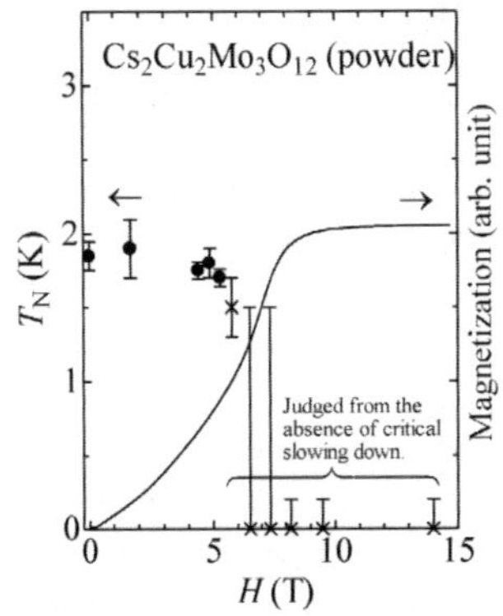


**Figure 6.** Field dependence of magnetization and $T_N$, estimated by NMR line width and $T_1^{-1}$.

Fujimura *et al.* [12], and hence confirms microscopically the existence of long range magnetic order at zero field. In general, however, the dynamical spin fluctuation contributes to $\lambda$ through its fluctuation amplitude $\delta H$ and its characteristic frequency $\omega_C$ as $\delta H^2/\omega_C$ [16]. So, our observation only assures an increase in $\delta H$ and/or a slowdown in $\omega_C$. We can assign it to the evidence of magnetic order under zero-field, because it agreed with the result of specific heat [12], and also with extrapolation of $T_N$ under finite fields.

We combine all $T_N$'s determined under various fields and plot them onto *H-T* diagram in Fig. 6. As increasing, field from zero, one can see that $T_N$ stays nearly constant until 5 T, from which it decreases steeply to zero at around 6 T, which is significantly lower than the saturation field $H_s = 9.15$ T [12]. In many cases of gapped quantum spin systems so far reported [19,20], a field-induced magnetic order takes place at finite temperature within a magnetization slope region, that is, the field at which $T_N$ becomes zero must coincide with $H_S$. The discrepancy found in this system suggests the relevancy with the spin nematic state under high fields [4-8]. However, in order to judge it, more detailed data are necessary, and furthermore, preparation of a single crystal is also required to avoid an ambiguity in the determination of $H_S$.

At the highest measured field 14.1 T, $1/T_1$ showed the thermal-activation type temperature dependence (Fig. 5), indicating opening of an energy gap in the spin-excitation spectrum. Adapting the value of saturation field $H_S = 9.15$ T [12], we compared the obtained gap $\Delta \simeq 7(1)$ K with the Zeeman energy $g\frac{1}{2}\mu_B(14.1 - H_s)$ to find that $g \simeq 4$. This anomalous value suggests a coherent flip of each neighboring two spins in the saturated field region [21], reflecting the ferromagnetic interaction between neighboring two spins.

Finally, we refer the temperature dependence of $1/T_1$ in the paramagnetic state. As shown in Fig. 4 (left), $1/T_1$'s for the entire field range below $H_s$ obey the power law above 5 K, which is shown by dashed curves. The power index shows significant dependence on the applied field. In the TLL state, the spatial spin correlation function obeys the power law as $x^{-2K}$, where the field-dependent constant $K$ is Luttinger parameter. The longitudinal and transverse components of spin fluctuation contribute to $1/T_1$ through the anisotropic hyperfine interaction, and let it exhibit the power law temperature dependence of $T^{2K-1}$ and $T^{1/2K-1}$, respectively [22-24]. In nematic or multipolar TLL state, where the transverse component is strongly suppressed at low temperatures [18,25], the term $T^{2K-1}$ solely dominates the temperature dependence of $1/T_1$. This means that $1/T_1$ may decrease at low temperatures when $K > \frac{1}{2}$. Note that in the ordinary TL liquid state [26,27], $1/T_1$ must always increase at low temperatures except for the singular point $K = \frac{1}{2}$. The observed behavior of $1/T_1$, that is, $2K - 1$ monotonically increases with increasing field as shown in Fig. 4 (left), seems to be well described above scenario. However, one must be very careful about the fact that $1/T_1$ generally decreases at low temperature in the saturation region, reflecting an opening of spin excitation gap. In order to conclude that the observed behavior is due to the characteristic spin fluctuation in the nematic TLL, determination of the hyperfine coupling tensor is indispensable, which are in the progress.

**Summary**

NMR and ZF μ-SR experiments on the $S = ½$ Heisenberg competing spin chain system $Cs_2Cu_2Mo_3O_{12}$ were performed in the wide-field range from zero to 14 T. The existence of long range magnetic order under the fields below 6 T was demonstrated by the broadening in NMR resonance line, critical divergence of $1/T_1$, or abrupt increase in μ-SR depolarization rate. $T_N \simeq 1.8$ K was nearly field independent until 6 T, above which $1/T_1$ decreased with decreasing temperature, indicating no magnetic order .

**Acknowledgement**

This work was partly supported by JSPS KAKENHI Grant Number 26108713, 15K05148 and. It was also supported by the Sophia University Special Grant for Academic Research. A part of the experiments was performed at High Field Laboratory for Superconducting Materials, Institute for Materials Research, Tohoku University.

**References**

[1] M. Hase *et al.,* J. Appl. Phys. 97 (2005) 10B303.
[2] M. Hase *et al.,* Phys. Rev. B70 (2004) 104426.
[3] S. F. Solodovnikov et al.: J. Struc. Chem. 38 (1997) 765.
[4] M. E. Zhitomirsky et al.: Euro Phys. Lett. 92 (2010) 37001.
[5] M. Hagiwara et al.: J. Phys. Conf. Ser. 320 (2011) 012049.
[6] H. Tsunetsugu et al.: J. Phys. Soc. Jap. 75 (2006) 083701.
[7] Lars Kecke et al.: Phys. Rev. B76, 060407R (2007).
[8] T. Hikihara et al.: Phys. Rev. B78 (2008) 144404.
[9] Y. Hoshino *et al.,* JPS Conf. Proc. 3 (2014) 014012.
[10] M. Urano *et al.*, Phys. Rev. B76 (2007) 024505.
[11] T. Suzuki et al.: J. Phys. Soc. Jpn. 76 (2007) 074704.
[12] A. Fujimura *et al.,* IEEE Transactions on Magnetics 99 (2016), *in press.*
[13] H. Inoue *et al.* , Phys. Rev. B79 (2009) 174418.
[14] T. Suzuki et al.: Phys. Rev. B79 (2009) 104409.
[15] S. Mishima et al.: J. Appl. Phys. 87 (2000) 5134.
[16] S. Nakajima *et al.,* J. Phys. Soc. Jpn. 81 (2012) 063706.
[17] K. Kanada *et al.*, J. Phys. Soc. Jpn. 76 (2007) 064706.
[18] M. Sato et al.: Phys. Rev. Lett. 110, 077206 (2013).
[19] G. Chaboussant et al.: Phys. Rev. Lett. 80 (1998) 2713.
[20] M. Jaime et al.: Phys. Rev. Lett. 93 (2004) 087203.
[21] N. Büttgen *et al.*: Phys. Rev. B90 (2014) 134401.
[22] T. Sakai, J. Phys. Soc. Jpn. 64 (1995) 251.
[23] T. Suzuki *et al.*: Phys. Rev. B 74 (2006) 172410.
[24] T. Suzuki *et al.*: Phys. Rev. B70 (2004) 054419.
[25] M. Sato et al. Phys. Rev. B83 (2011) 064405.
[26] M. Klanšjek et al.: Phys. Rev. Lett. 101 (2008) 13727.
[27] K. Ninios et al.: Phys. Rev. Lett. 108 (2012) 097201.

IOP Conf. Series: Journal of Physics: Conf. Series **828** (2017) 012018 doi:10.1088/1742-6596/828/1/012018

# Neutron Scattering Study in Breathing Pyrochlore Antiferromagnet $Ba_3Yb_2Zn_5O_{11}$

**T. Haku[1], M. Soda[1], M. Sera[2], K. Kimura[2], J. Taylor[3], S. Itoh[4], T. Yokoo[4], Y. Matsumoto[1], D. Yu[5], R. A. Mole[5], T. Takeuchi[6], S. Nakatsuji[1], Y. Kono[1], T. Sakakibara[1], L.-J. Chang[7],T. Masuda[1]**

[1]Institute for Solid State Physics, The University of Tokyo, Kashiwanoha, Kashiwa, Chiba 277-8581, Japan

[2]Division of Materials Physics, Graduate School of Engineering Science, Osaka University, Toyonaka, Osaka 560-8531, Japan

[3]ISIS Facility, Rutherford Appleton Laboratory, Chilton, Didcot, OX11 0QX, United Kingdom

[4]Institute of Materials Structure Science, High Energy Accelerator Research Organization, Tsukuba, Ibaraki 305-0801, Japan

[5]Bragg Institute, Australian Nuclear Science and Technology Organization, Lucas Heights, New South Wales 2234, Australia

[6]Low Temperature Center, Osaka University, Toyonaka, Osaka 560-0043, Japan

[7]Department of Physics, National Cheng Kung University, Tainan 70101, Taiwan

E-mail: masuda@issp.u-tokyo.ac.jp

**Abstract.** Inelastic neutron scattering (INS) study on breathing pyrochlore antiferromagnet $Ba_3Yb_2Zn_5O_{11}$ is presented. Observed crystalline electric field (CEF) excitations are explained by a Hamiltonian of Kramers ion $Yb^{3+}$ of which the local symmetry exhibits $C_{3v}$ point group symmetry. The magnetic susceptibility is consistently reproduced by the energy scheme of the CEF excitations. The INS spectra in the low-energy range are quantitatively explained by spin-1/2 single-tetrahedron model having *XXZ* anisotropy and Dzyaloshinskii-Moriya interaction. This model has a two-fold degeneracy of the lowest-energy state per tetrahedron and well reproduces the bulk properties at $T \geq 0.5$K. At lower temperatures, however, we observe a broad maximum in the heat capacity around 63 mK, demonstrating that a unique quantum ground state is selected due to extra perturbations.

Geometrical frustrated magnets have attracted great attention because they exhibit variety of novel states [1, 2]. Among the frustrated lattices a three-dimensional network of corner-sharing tetrahedra, i.e., the pyrochlore lattice, is one of the most interesting systems. The geometrical configuration prohibits the arrangement of the spins that satisfy the lowest energy of all spin bonds and it leads to the ground state sensitive to perturbations. Indeed various states have been experimentally studied in rare-earth pyrochlore compounds $R_2Ti_2O_7$ [3]. Recently breathing pyrochlore (BP) lattice, i.e., alternating array of corner-sharing large and small tetrahedra, has been focused in terms of a prototype of the pyrochlore lattice. Exotic states including a quantum spin liquid [4], an ordering of spin singlet state [5, 6], and a chiral order state [7, 8] have been predicted. As model compounds $S = 3/2$ spinel compounds $LiInCr_4O_8$ and $LiGaCr_4O_8$ were reported [9, 10]. They, however, exhibited a magnetostructural transition and the ground states are ordered states.

Recently a novel candidate for the BP lattice, $Ba_3Yb_2Zn_5O_{11}$, was reported [11]. Kramers ions $Yb^{3+}$ carrying effective $S = 1/2$ spins form regular tetrahedra, small one with the intratetrahedra distance of 3.3 Å and the large one with that of 6.2 Å. The $Yb^{3+}$ ion is surrounded by six oxygen ions and the point symmetry of the $YbO_6$ octahedron is $C_{3v}$. No phase transition were observed in the specific heat measurement in the temperature range 0.4 K $< T <$ 300 K. The bulk magnetic properties in the low temperatures were explained by Heisenberg spin $S = 1/2$ tetrahedron model having non-magnetic ground state. For the further understanding of the spin model and its ground state, inelastic neutron scattering experiment is necessary. In this proceedings we review our recent neutron studies on $Ba_3Yb_2Zn_5O_{11}$ [12, 13].

The inelastic neutron scattering (INS) experiments were performed at three different spectrometers; MARI in ISIS, HRC in J-PARC, and PELICAN in ANSTO. In the MARI experiment, we used Gd chopper with 100Hz for choosing the incident energy of neutron $E_i$ of 6 meV and that with 350 Hz for $E_i$ of 150 meV. Combination of closed-cycle refrigerator and 1 K insert was used to achieve 2.0 K. We used 17.7 g powder sample synthesized by a solid state reaction method. The details for HRC and PELICAN experiments were described in Refs. [12] and [13], respectively.

Figure 1 shows INS spectrum measured using MARI spectrometer with $E_i$ = 150 meV at 5.0 K. Three flat excitations are clearly observed at $\hbar\omega$ = 38.2, 55.0, and 68.3 meV, where $\hbar\omega$ is the energy transfer of neutrons. With the increase of $Q$ in the range of $Q \leq 6$ Å$^{-1}$ the intensities decrease, where $Q$ is scaler of the wave vector transfer. This means that the excitations are from magnetic clusters, and, therefore, they are regarded as transitions between energy levels of $J$ multiplets lifted by crystalline electric field (CEF) of the magnetic Yb ions. In the $C_{3v}$ symmetry the 8-fold multiplets of Yb ion is lifted into four Kramers doublets and it is observed in the INS spectrum. The first excited energy is 38.2 meV that is higher than room temperature, and thus the Yb ion is regarded as effective spin $S = 1/2$ at the lower temperatures. Excitations smeared in the low energy and high $Q$ region are regarded as phonon excitations. In the HRC experiment [12] consistent results were obtained and the CEF excitations were analyzed on the basis of the Hamiltonian,

$$\mathcal{H}_{CEF} = B_2^0 O_2^0 + B_4^0 O_4^0 + B_4^3 O_4^3 + B_6^0 O_6^0 + B_6^3 O_6^3 + B_6^6 O_6^6, \tag{1}$$

where $B_m^n$ and $O_m^n$ are CEF parameters and Steven's operators, respectively [14, 15]. The obtained parameters are summarized in Table 1. Using the parameters $B_m^n$ the magnetic susceptibility [11] was reasonably reproduced [12]. The wave function of the ground state suggested that the effective spin had easy-plane anisotropy.

**Table 1.** Values for $B_n^m$ of $Ba_3Yb_2Zn_5O_{11}$ obtained by INS experiment.

| | $B_2^0$ | $B_4^0 \times 10^2$ | $B_4^3 \times 10^2$ | $B_6^0 \times 10^4$ | $B_6^3 \times 10^4$ | $B_6^6 \times 10^4$ |
|---|---|---|---|---|---|---|
| Energy (meV) | 0.0737 | −2.76 | −33.6 | 6.10 | 8.70 | 88.2 |

Figure 2(a) and 2(b) show the INS spectrum measured with $E_i$ = 6 meV at $T$ = 2.0 K and 30 K, respectively. At 2.0 K flat excitations are observed at $\hbar\omega \sim$ 0.4, 0.5 and 0.8 meV. At 30 K additional ones are observed at 1.0, 1.3, and 1.8 meV at 30 K. All the excitations have the maximum intensities at $Q \sim$ 1.2 Å$^{-1}$, which corresponds to the inverse length scale of intratetrahedron of Yb ions, and the intensities decrease with the increase of $Q$. They are, thus, regarded as excitations of spin $S = 1/2$ tetrahedra of Yb ions. In contrast with the experimental data exhibiting many modes, single excitation is theoretically expected in Heisenberg spin tetrahedron at zero temperature. For the further analysis, combination of high resolution data and the Hamiltonian including anisotropic and/or asymmetric interactions is necessary. Then, we collected high resolution data in the low energy range using PELICAN spectrometer in ANSTO [13]. The obtained data were consistent with those obtained in MARI.

For the analysis of the PELICAN data we considered the following general expression[16]:

$$\mathcal{H} = -\sum_{i<j}\sum_{\nu\mu} J_{ij}^{\nu\mu} S_i^{\nu} S_j^{\mu}. \tag{2}$$

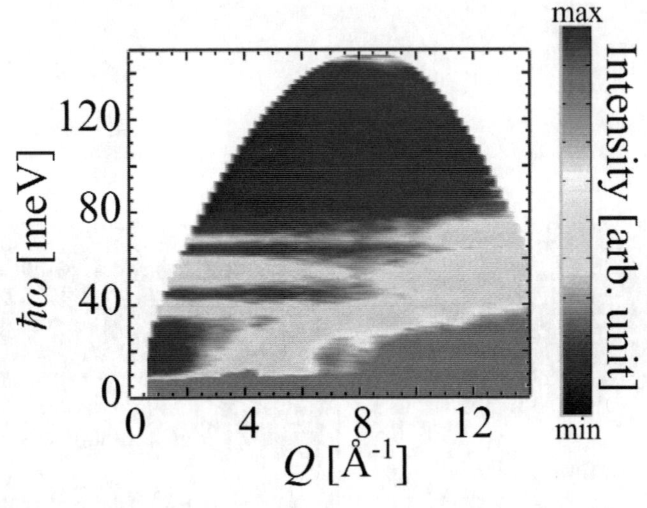


**Figure 1.** INS spectrum of $Ba_3Yb_2Zn_5O_{11}$ measured with $E_i = 150$ meV at $T = 5.0$ K at MARI spectrometer.

Here $i$ and $j$ are the labels of the spins on the tetrahedron and $\nu$ and $\mu$ represent Cartesian coordinates $x$, $y$, and $z$ which are defined along the crystallographic $a$, $b$ and $c$-axes, respectively. The position vectors for the spins are $\boldsymbol{r}_1 = d/\sqrt{3}(1,1,1)$, $\boldsymbol{r}_2 = d/\sqrt{3}(-1,-1,1)$, $\boldsymbol{r}_3 = d/\sqrt{3}(1,-1,-1)$, and $\boldsymbol{r}_4 = d/\sqrt{3}(-1,1,-1)$, where $2\sqrt{2}d$ is the length of the side of the tetrahedron. The symmetry of the regular tetrahedron determines the form of the interaction tensor $\hat{J}_{ij}$[16], for example in the case of $\hat{J}_{12}$ one gets,

$$\hat{J}_{12} = \begin{pmatrix} J_{xx} & J_{yx} & J_{zx} \\ J_{xy} & J_{yy} & J_{zy} \\ J_{xz} & J_{yz} & J_{zz} \end{pmatrix} = \begin{pmatrix} J_1 & J_3 & -J_4 \\ J_3 & J_1 & -J_4 \\ J_4 & J_4 & J_2 \end{pmatrix}. \tag{3}$$

Using the Hamiltonian in Eq. (2) and the interaction tensor $\hat{J}_{ij}$, we calculated the neutron scattering cross section. We performed a numerical fitting to the INS spectra measured at $T = 1.5$ K using a genetic algorithm. The fitting parameters were $J_1$, $J_2$, $J_3$, $J_4$ and a scale factor. Here $J_1$ is an $XY$ type interaction, $J_2$ is Ising type, $J_3$ is a pseudo-dipole-type and $J_4$ is Dzyaloshinskii-Moriya (DM) type. The magnitude of the Dzyaloshinskii vector correspond to $\sqrt{2}J_4$. The best fit parameters were:

$$J_1 = -0.570 \pm 0.033 \text{ meV}, \tag{4}$$
$$J_2 = -0.558 \pm 0.028 \text{ meV}, \tag{5}$$
$$J_3 = 0.000 \pm 0.023 \text{ meV}, \tag{6}$$
$$J_4 = 0.113 \pm 0.014 \text{ meV}. \tag{7}$$

These were consistent within error to those obtained from independent INS experiments[17]. The results showed that the system had an easy-plane $XXZ$ type anisotropy with the DM-type interaction about 30 % of the main interaction. INS spectra measured at various temperatures were completely reproduced by the parameters in Eqs. (4)-(7). Thus we concluded that the neutron data were explained by $S = 1/2$ spin tetrahedron including DM interaction in the temperature range $1.5\text{ K} \leq T \leq 40$ K.

In Ref. [13] heat capacity at 0 T in the temperature range $0.5\text{ K} \leq T \leq 15$ K and magnetization at 0.5 K in the field range $0\text{ T} \leq H \leq 12.5$ T were calculated using Eqs. (2) and (4)-(7), and the experimental data were reasonably reproduced. The ground state of the obtained spin Hamiltonian was, however,

IOP Conf. Series: Journal of Physics: Conf. Series **828** (2017) 012018 doi:10.1088/1742-6596/828/1/012018

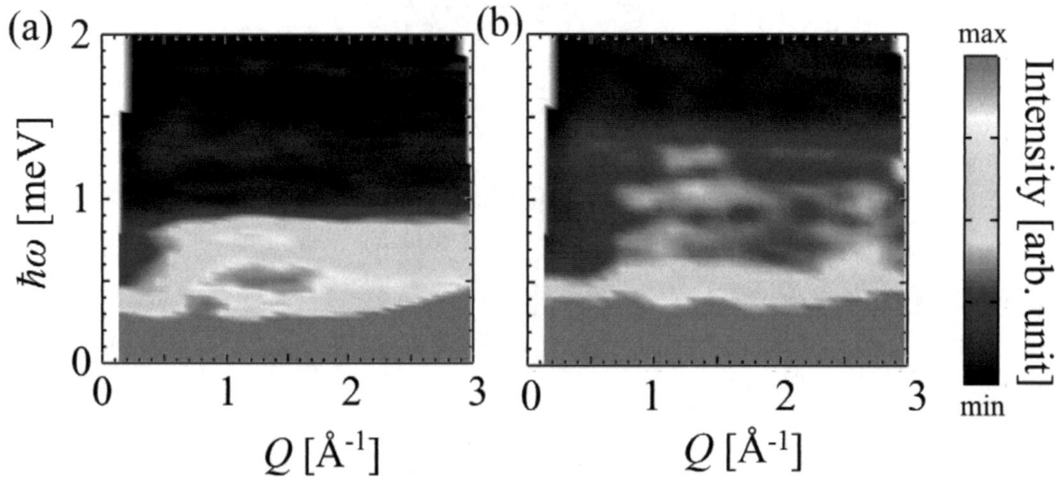


**Figure 2.** INS spectrum of $Ba_3Yb_2Zn_5O_{11}$ measured with $E_i$ = 6 meV at $T$ = 2.0 K (a) and 30 K (b) at MARI spectrometer.

doublet and a unique state should be selected at very low temperature from the requirement of the third law of thermodynamics. By measuring the heat capacity down to 30 mK, absence of jump was confirmed and broad maximum at 63 mK was observed. The change of the entropy was consistent to two degrees of freedom per Yb tetrahedron, meaning that a unique ground state was selected due to extra perturbations with an energy scale smaller than the energy resolution of the INS experiment. Since the state was free from magnetic order, it was a novel spin liquid. Intertetrahedra interaction and spin-lattice interaction were discussed as the perturbation in Ref. [13].

Combination of neutron scattering experiment in high energy and low energy regions identified the spin Hamiltonian of the breathing pyrochlore antiferromagnet $Ba_3Yb_2Zn_5O_{11}$. Heat capacity measurement down to ultralow temperature revealed the realization of spin liquid state in this compound.

**Acknowledgments**
Travel expenses for the neutron scattering experiment performed using MARI at ISIS, United Kingdom, were supported by the General User Program for Neutron Scattering Experiments, Institute for Solid State Physics, The University of Tokyo (Proposal No. 14522 ), at JRR-3, Japan Atomic Energy Agency, Tokai, Japan.

**References**
[1] A. P. Ramirez, Annu. Rev. Mater. Sci. **24**, 453 (1994).
[2] L. Balents, Nature **464**, 199 (2010).
[3] J. S. Gardner, M. J. P. Gingras, and J. E. Greedan, Rev. Mod. Phys. **82**, 53 (2010).
[4] B. Canals and C. Lacroix, Phys. Rev. Lett. **80**, 2933 (1998).
[5] H. Tsunetsugu, Phys. Rev. B **65**, 024415 (2001).

[6] E. Berg, E. Altman, and A. Auerbach, Phys. Rev. Lett. **90**, 147204 (2003).
[7] J. H. Kim and J. H. Han, Phys. Rev. B **78**, 180410(R) (2008).
[8] F. J. Burnell, S. Chakravarty, and S. L. Sondhi, Phys. Rev. B **79**, 144432 (2009).
[9] Y. Okamoto, G. J. Nilsen, J. P. Attfield, and Z. Hiroi, Phys. Rev. Lett. **110**, 097203 (2013).
[10] G. J. Nilsen, Yoshihiko Okamoto, Takatsugu Masuda, Juan Rodriguez-Carvajal, Hannu Mutka, Thomas Hansen, and Zenji Hiroi, Phys. Rev. B **91**, 174435 (2015).
[11] K. Kimura, S. Nakatsuji, and T. Kimura, Phys. Rev. B **90**, 060414(R) (2014).
[12] T. Haku, M. Soda, M. Sera, K. Kimura, S. Itoh, T. Yokoo, and T. Masuda, Journal of the Physical Society of Japan **85**, 034721 (2016).
[13] T. Haku, K. Kimura, Y. Matsumoto, M. Soda, M. Sera, D. Yu, R.A. Mole, T. Takeuchi, S. Nakatsuji, Y. Kono, T. Sakakibara, L.-J. Chang, and T. Masuda, Phys. Rev. B **93**, 220407(R) (2016).
[14] M. J. Hutchings, Solid State Phys. **16**, 227 (1964).
[15] K. W. H. Stevens, Proc. Phys. Soc., Sect. A Proc. **65**, 209 (1952).
[16] K. A. Ross, L. Savary, B. D. Gaulin, and L. Balents, Phys. Rev. X **1**, 021002 (2011).
[17] J.G. Rau, L.S. Wu, A.F. May, L. Poudel, B. Winn, V.O. Garlea, A. Huq, P. Whitfield, A.E. Taylor, M.D. Lumsden, M.J.P. Gingras, and A.D. Christianson, Phys. Rev. Lett. **116**, 257204 (2016).

# DM-induced frustration of the weakly coupled Heisenberg chains

**Wen Jin and Oleg A. Starykh**

Department of Physics and Astronomy, University of Utah, Salt Lake City, Utah 84112, USA

E-mail: wen.jin@utah.edu

**Abstract.** We present theoretical procedure for estimating interchain exchange coupling $J'$ between antiferromagnetic spin-1/2 Heisenberg chains with frustration due to Dzyaloshinskii-Moriya (DM) interaction characterized by DM vectors $\mathbf{D}_y$ which are uniform within each chain $y$, but staggered between adjacent chains, $\mathbf{D}_y \sim (-1)^y \mathbf{D}$. Under a magnetic field $\mathbf{h} \parallel \mathbf{D}$ we obtain a field-temperature phase diagram which favorably agrees with the one experimentally observed. We then apply chain mean-field (CMF) technique to calculate interchain exchange $J'$ from the critical field $h_c$ at which the transition between the collinear spin-density wave and the cone states takes place. The CMF calculations are found to provide good physical description of the experimental measurements for the wide range of $D/J'$.

## 1. Introduction

Many experimental realizations of Heisenberg spin chain materials have become available over recent years. Here we present comparative analysis of two new interesting materials – $K_2CuSO_4Cl_2$ and $K_2CuSO_4Br_2$ [1, 2] – which represent an interesting and novel case of Heisenberg spin chains with *uniform* Dzyaloshinskii-Moriya (DM) interactions.

Both materials are believed to be described by the Hamiltonian (1) below and are characterized by a different set of parameters $(J, D, J')$, where $J$ is the dominant intra-chain exchange, $D$ is the DM interaction strength and $J'$ is the interchain spin exchange. Despite close structural similarity, the two materials are characterized by different $h - T$ phase diagrams, as has been established by experiments in Prof. Zheludev's group in ETH [1–3]. That difference is attributed to the different $D/J'$ ratio – according to Ref. [1] DM interaction is relatively weak in $K_2CuSO_4Cl_2$, so that $D/J' \sim 1$, while $K_2CuSO_4Br_2$ can be characterized as a strong DM material with $D/J' \sim 10$.

As we describe below and show in more details elsewhere [4], large $D/J'$ ratio places Br-based material into a novel category of materials where interchain interaction between spins from adjacent chains is strongly frustrated by the uniform in-chain, but staggered between chains, DM interaction. This unique geometry of DM interactions makes $K_2CuSO_4Br_2$ somewhat similar to the honeycomb iridates family $Li_2IrO_3$ an incommensurate magnetic order of which is characterized by unusual counter-rotating spirals on neighboring sublattices [5]. This kind of frustration requires theoretical re-evaluation of the kind and mechanism of the eventual two- (or, three-) dimensional magnetic order that develops in the system at sufficiently low temperature. Our work provides such an analysis.

IOP Conf. Series: Journal of Physics: Conf. Series **828** (2017) 012019 doi:10.1088/1742-6596/828/1/012019

A variety of experimental techniques has been employed to characterize the parameters of $K_2CuSO_4Cl_2$ and $K_2CuSO_4Br_2$ [1, 2]. The dominant intra-chain exchange $J$ has been estimated using the empirical fitting function of Ref. [6] to fit the uniform magnetic susceptibility data as well as by fitting the inelastic neutron scattering continuum, a unique feature of the Heisenberg spin-1/2 chain, to the Müller ansatz [7]. DM vector $\mathbf{D}$ has been measured by electron spin resonance (ESR) [2, 8]. However the interchain exchange interaction $J'$ has been estimated from the chain mean-field theory fit based on Monte-Carlo improved study in Ref. [9]. This fit, however, completely neglects crucial for understanding of these materials DM interactions and moreover assumes that spin chains form simple non-frustrated cubic structure. The second assumption is not justified as well. Inelastic neutron scattering data show that the interchain exchange between spin chains in the $a-b$ plane is at least an order of magnitude stronger than that along the $c$-axis, connecting different $a-b$ planes. As a result, it is more appropriate to consider the current problem as two-dimensional whereby spin chains, running along the $a$-axis, interact weakly via $J' \ll J$ directed along the $b$-axis.

This is the geometry assumed in the present work. Using bosonization technique, we account for the intra-chain DM interaction non-perturbatively, and describe the competition between the cone and the longitudinal spin-density wave (SDW) orders. The interchain $J'$ is estimated from the value of the zero-field critical temperature $T_c$, which is calculated with the help of the chain mean field (CMF) approximation [10]. The obtained $T_c$'s are found to be in a good agreement with experimental measurements.

## 2. Hamiltonian

We consider weakly coupled antiferromagnetic Heisenberg spin-1/2 chains subject to a uniform Dzyaloshinskii-Moriya (DM) interaction and an external magnetic field. The system is described by the following Hamiltonian,

$$\mathcal{H} = \sum_{x,y}[J\mathbf{S}_{x,y} \cdot \mathbf{S}_{x+1,y} + J'\mathbf{S}_{x,y} \cdot \mathbf{S}_{x,y+1}] + \mathbf{D} \cdot \sum_{x,y}(-1)^y \mathbf{S}_{x,y} \times \mathbf{S}_{x+1,y} - h \cdot \sum_{x,y} S^z_{x,y}, \quad (1)$$

where $\mathbf{S}_{x,y}$ is the spin-1/2 operator at site $(x, y)$. $J$ and $J'$ denote isotropic intra- and inter-chain antiferromagnetic exchange couplings as shown in Fig. 1, and we account for interactions between nearest neighbors only. The inter-chain exchange is weak, of the order of $J' \sim 10^{-2}J$. DM interaction is parameterized by the DM vector $\mathbf{D} = D\hat{z}$, direction of which is staggered (note the factor $(-1)^y$) between adjacent chains. Importantly, vector $\mathbf{D}$ is uniform within a given $y$-th chain. $h = g\mu_B B$ is an external magnetic field along DM interaction.

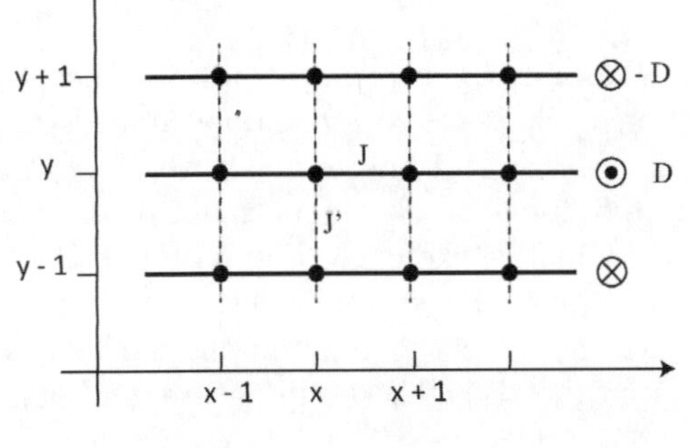


**Figure 1.** Sites, exchange couplings and DM vectors on coupled spin chains. Intra-chain bonds $J$ (thick lines along $\hat{x}$), inter-chain bonds $J'$ (dashed lines along $\hat{y}$), and $J' \ll J$. DM vectors on neighboring chain have opposite direction, either point into or out off paper.

### 2.1. Bosonization

In the low-energy continuum limit, Hamiltonian is expressed by bosonization [11–13],

$$\mathcal{H}_{\text{chain}} = \tilde{\mathcal{H}}_0 + \tilde{\mathcal{H}}_{\text{bs}} + \tilde{\mathcal{H}}_{\text{inter}}, \quad (2)$$

where $\tilde{\mathcal{H}}_0$ has quadratic form in terms of abelian bosonic fields $\phi, \theta$ (see details in Appendix A), the Zeeman and DM interaction terms are absorbed in $\tilde{\mathcal{H}}_0$ by linear shifts of fields $\phi$ and $\theta$, correspondingly. The harmonic Hamiltonian is perturbed by chain backscattering $\tilde{\mathcal{H}}_{\rm bs}$ and inter-chain $\tilde{\mathcal{H}}_{\rm inter}$ interactions, which read

$$\tilde{\mathcal{H}}_{\rm bs} = \int dx \Big\{ \pi v y_B (J_R^+ J_L^- e^{-i2t_\phi x} + {\rm h.c.}) + 2\pi v y_z J_R^z J_R^z \Big\}, \; t_\phi \equiv h/v, \tag{3}$$

and $\tilde{\mathcal{H}}_{\rm inter} = \mathcal{H}_{\rm cone} + \mathcal{H}_{\rm sdw}$, where

$$\mathcal{H}_{\rm cone} = c_1 \int dx \cos[\beta(\theta_y - \theta_{y+1}) + 2(-1)^y t_\theta x], \quad \mathcal{H}_{\rm sdw} = c_2 \int dx \cos[\frac{2\pi}{\beta}(\phi_y - \phi_{y+1})]. \tag{4}$$

The coupling constants are,

$$c_1 = J' A_3^2, \quad c_2 = J' A_1^2/2. \tag{5}$$

$v \simeq J\pi a/2$ is the spin velocity, $a$ is lattice constant and $t_\theta \equiv D/v$. $\mathbf{J}_L(x)$ and $\mathbf{J}_R(x)$, are the uniform left and right spin currents, which are defined in Appendix A. The parameter $\beta = 2\pi R$ is related to the "compactification radius" $R$ in the sine-Gordon (SG) model. In the absence of external field, the SU(2) invariant Heisenberg chain has $2\pi R^2 = 1$. The amplitudes $A_1$ and $A_3$ have been determined numerically [14]. $\mathcal{H}_{\rm cone}$ and $\mathcal{H}_{\rm sdw}$ are the transverse and longitudinal (with respect to the $z$-axis) components of inter-chain interaction respectively.

**Table 1.** Exchange constants for $K_2CuSO_4Cl_2$ and $K_2CuSO_4Br_2$: The intrachain exchange $J$ is obtained by thermodynamic and neutron measurements. $D$ values are from ESR measurements. Inter-chain exchange $J'$ in fourth column is obtained by fitting data to Ref. [9]; inter-chain exchange $\mathcal{J}'$ in the last column is obtained by fitting data to our CMF calculation in Eq. (7), (9) and (10).

| | $J$ | $D$ | $J'$ by Ref. [9] | $\mathcal{J}'$ by CMF |
|---|---|---|---|---|
| $K_2CuSO_4Cl_2$ | 3.1 K | $\sim$ 0.04 K | 0.031 K | 0.073 K |
| $K_2CuSO_4Br_2$ | 20.5 K | 0.28 K | 0.034 K | 0.20 K |

### *2.2. Two phase diagrams*

We notice that there is a position dependent oscillation term in $\mathcal{H}_{\rm cone}$ (4). The physical meaning is that staggered DM interaction forces spins in neighboring chains to rotate in opposite direction thereby frustrating the transverse inter-chain interaction. If the DM interaction is weak, $D \ll J'$ (the case of $K_2CuSO_4Cl_2$), oscillation in $\mathcal{H}_{\rm cone}$ is slow and does not affect renormalization group (RG) flow of the coupling constant $c_1$. One finds that in the presence of external magnetic field the more relevant $\mathcal{H}_{\rm cone}$ dominates over the $\mathcal{H}_{\rm sdw}$ and the ground state is cone state. Minimization of the argument of cosine in $\mathcal{H}_{\rm cone}$ has the effect of undoing the shift (A.9), resulting in a *commensurate cone* configuration shown in Fig. B1. The temperature-field phase diagram for this case is shown in Fig. 3.

If the DM interaction is strong, $D \gg J'$ (the case of $K_2CuSO_4Br_2$), $\mathcal{H}_{\rm cone}$ oscillates rapidly and averages to zero. As a result, the only inter-chain interaction that survives in this situation is the $\mathcal{H}_{\rm sdw}$, which promotes incommensurate longitudinal SDW order. However, a cone-like interaction between more distant chains can be generated by quantum fluctuations at low energies [15]. The simplest of such interactions is given by the transverse interaction between the *next-neighbor* (NN) chains $\mathcal{H}_{\rm NN}$ (derivation of this term is presented in Ref. [4]),

$$\mathcal{H}_{\rm NN} = -c_3 \int dx \cos[\beta(\theta_y - \theta_{y+2})], \quad c_3 = \frac{\pi}{4}\frac{J'^2}{D} A_3^4 t_\theta^{2\Delta_1 - 1} \frac{\Gamma(1-\Delta_1)}{\Gamma(\Delta_1)}, \tag{6}$$

where $\Delta_1 \simeq 1/2$ at low fields. This is an indirect exchange, mediated by an intermediate chain $(y+1)$, and therefore its exchange coupling is rather weak, $(J')^2/D \ll J'$. Importantly this interaction is not frustrated (spins in chains $y$ and $y+2$ are rotating in the *same* direction) and becomes more relevant with increasing magnetic field. $\mathcal{H}_{\rm NN}$ competes with $\mathcal{H}_{\rm sdw}$, and results in a phase transition from SDW to coneNN, a *cone-like incommensurate order* induced by $\mathcal{H}_{\rm NN}$, at some critical field $h_c$. For $K_2CuSO_4Br_2$ it appears that $h_c \approx 0.1$ T [1]. This coneNN order corresponds to $\theta_y - \theta_{y+2} = 2\pi/\beta$, and the resulting spiral spin configuration is illustrated in Fig. 2. The main role of DM interaction is to induce incommensurability, $\propto D/v$, in the *counter-rotating* spiral orderings in the neighboring spin chains, see Appendix B for more details.

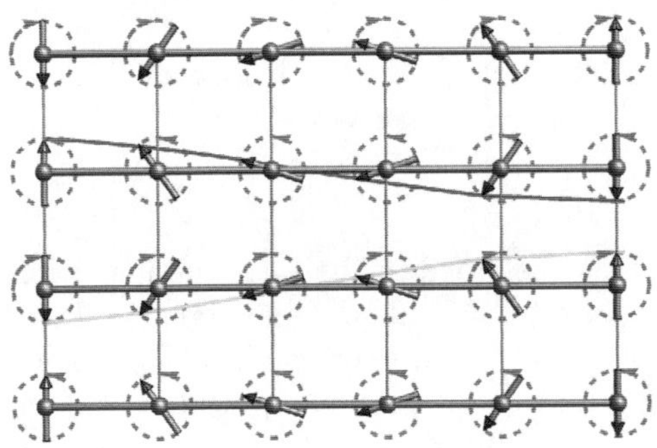

**Figure 2.** Staggered magnetization of the coneNN state, illustrated for a field $\mathbf{h} \parallel \mathbf{D}$, where all spins are ordered in the transverse plane. Red circles with arrows indicates the procession direction of spins, as one moves along each chain. Note that the arrows' direction alternates between consecutive chains, owing to the staggering of the DM vector. Blue and green curves illustrate spin orientation in neighboring chains.

## 3. Chain mean field calculation

The chain mean-field (CMF) approximation [10] allows one to calculate critical temperatures for different possible instabilities desribed in Sec. 2.2. There are three states that need to be considered, and we present the critical temperature for each state in the following. Details of the calculations can be found in Ref. [10] and Appendix C.

### 3.1. Cone order

Ordering temperature for the cone state $T_{\rm cone}$ is the solution of equation,

$$1 = \eta_1 (2\pi T_{\rm cone}/v)^{2\Delta_1 - 2} \frac{\Gamma(1-\Delta_1)}{\Gamma(\Delta_1)} |\Gamma(\Delta_1/2 + iy)|^4 [\cosh(2\pi y) - \cos(\pi\Delta_1)], \tag{7}$$

with

$$y = \frac{t_\theta v}{4\pi T_{\rm cone}}, \quad \eta_1 = \frac{c_1}{2\pi v}. \tag{8}$$

$\Delta_1$ is the scaling dimension of $N^+$, its value in the limits of zero and full magnetization is shown in Table 2. The plot of $T_{\rm cone}$ for $K_2CuSO_4Cl_2$ is shown as the blue curve in Fig. 3 – this corresponds to Fig. 14 in Ref. [1]. Here we took the exchange constants $J = 3.1$ K and $D = 0.04$ K from Table 1. By fitting the zero-field experimental value of $T_{\rm cone}$ ($T_{\rm cone}|_{h=0}$ =77 mK in Ref. [1]) to Eq. (7), where $\Delta_1 = 1/2$, we obtain $\mathcal{J}'_{\rm Cl} = 0.073$ K. The (approximately) factor of 2 difference between our result and the previous estimate in Ref. [1] is caused by the

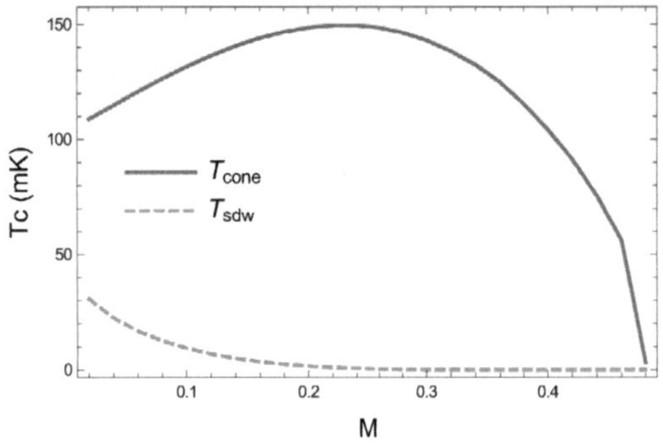


**Figure 3.** Critical temperatures of SDW ($T_{\rm sdw}$, orange dashed line) and cone ($T_{\rm cone}$, blue solid) as a function of magnetization $M$ for $K_2CuSO_4Cl_2$, with $J = 3.1$ K, $D = 0.04$ K in Table 1 and $\mathcal{J}'_{\rm Cl} = 0.073$ K from Eq. (7) by setting $T_{\rm cone}|_{h=0}$ to 77 mK. The order with the larger critical temperature is the one realized. Here the phase diagram consists of a single cone phase.

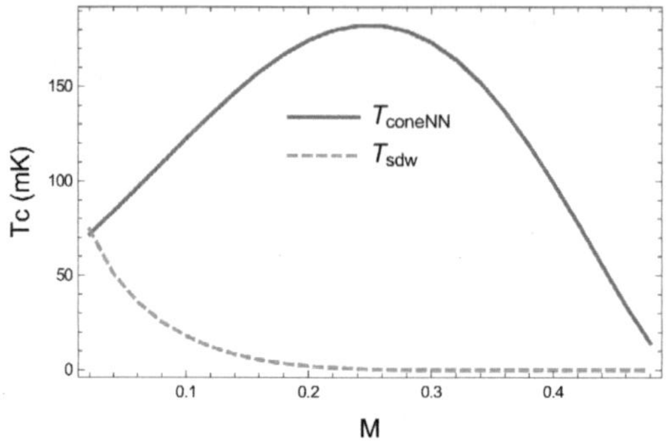


**Figure 4.** Critical temperatures of SDW ($T_{\rm sdw}$, orange dashed line) and coneNN ($T_{\rm coneNN}$, blue solid line) as a function of magnetization $M$ for $K_2CuSO_4Br_2$, with $J = 20.5$ K, $D = 0.28$ K are taken from Table 1 , and $\mathcal{J}'_{\rm Br} = 0.20$ K from critical field $B_c \sim 0.1$ T (details in Sec. 4). Two curves intersect at very low field, indicating a SDW-coneNN phase transition.

accepted here two-dimensional geometry of the system, as described in the Introduction. Fig. 3 shows that $T_{\rm cone}$ is enhanced by field, and has a max close to 150 mK, which agrees well with the experimental maximum of $T_{\rm cone}$, see Fig. 14 in Ref. [1]. At high magnetic field $T_{\rm cone}$ starts to decrease because orthogonal to the magnetic field spin projection (the amplitude $A_3$ in (5)) decreases to zero on approaching the fully polarized phase.

Importantly, increasing the ratio $D/J'$ suppresses $T_{\rm cone}$, until at some critical value $D_c/J' \approx 1.2$, the solution of Eq. (7) disappears completely. Mathematically, this value is determined by the maximum of the right-hand-side of Eq. (7) as a function of $y$ [4]. Physically, strong DM interaction ($D/J' \gg 1$) frustrates transverse interchain coupling, effectively turning it off. As a result, the cone order is destroyed.

### *3.2. SDW order*

The ordering temperature for SDW state is,

$$T_{\rm sdw} = \frac{v}{2\pi}\Big[\eta_2 \frac{\Gamma(1-\Delta_2)\Gamma(\Delta_2/2)^2}{\Gamma(\Delta_2)\Gamma(1-\Delta_2/2)^2}\Big/\Big(1+\eta_2\frac{\Gamma(\Delta_2-1/2)}{\sqrt{\pi}(1-\Delta_2)\Gamma(\Delta_2)}\Big)\Big]^{1/(2-2\Delta_2)}, \tag{9}$$

with $\eta_2 = \pi c_2/v$. The denominator compensates the non-physical divergence which occurs when $\Delta_2 \to 1$ near the saturation transition, see Table 2. Calculated $T_{\rm sdw}$ for $K_2CuSO_4Cl_2$ and $K_2CuSO_4Br_2$ are shown in orange in Fig. 3 and Fig. 4, respectively. In Fig. 3, $T_{\rm sdw}$ is smaller than $T_{\rm cone}$, so that SDW does not realize. However, in Fig. 4, $T_{\rm sdw}$ is dominant at small fields $h < h_c$. At higher $h$ (higher magnetization $M$) $T_{\rm coneNN}$ takes over, corresponding to the SDW-coneNN phase transition.

*3.3. ConeNN*

When it comes to coneNN state, its ordering temperature has a simple form, due to the fact that $\mathcal{H}_{\rm NN}$ is free from oscillation and $T_{\rm coneNN}$ is free from divergence ($\Delta_1 \leq 1/2$),

$$T_{\rm coneNN} = \frac{v}{2\pi}\Big[\eta_3 \frac{\Gamma(1-\Delta_1)\Gamma(\Delta_1/2)^2}{\Gamma(\Delta_1)\Gamma(1-\Delta_1/2)^2}\Big]^{1/(2-2\Delta_1)}, \tag{10}$$

with $\eta_3 = \pi c_3/v$. $T_{\rm coneNN}$ for $K_2CuSO_4Br_2$ is shown in Fig. 4, where values for $J$ and $D$ are from Table 1. $\mathcal{J}'_{\rm Br} = 0.20$ K is obtained from the calculation in Sec. 4, where we describe that for strong DM interaction the value of the interchain exchange $\mathcal{J}'$ can be obtained directly from the transition field $h_c$.

**Table 2.** Scaling dimensions $\Delta_{1/2}$ of transverse ($N^{\pm}$) and longitudinal ($N^z$) components of staggered magnetization $\mathbf{N}$ at magnetization $M$. The last two columns are the scaling dimensions in the limit of zero and full polarization, respectively.

| | Operator | Field | Expression | $M=0$ | $M=1/2$ |
|---|---|---|---|---|---|
| $\Delta_1$ | $N^{\pm}$ | $\theta$ | $\pi R^2$ | 1/2 | 1/4 |
| $\Delta_2$ | $N^z$ | $\phi$ | $\pi/\beta^2$ | 1/2 | 1 |

## 4. Determination of $\mathcal{J}'$ from the critical field $h_c$ for strong DM interaction

Experiments on $K_2CuSO_4Br_2$ have observed phase transition at a very small magnetic field $B_c \sim 0.1$ T [1, 2]. This corresponds to a critical field $h_c = 0.1343$ K (with $h_c = g\mu_B B_c$, and set $\mu_0 = 1$). Clearly $h_c \ll J$, which is consistent with our low-field consideration. Using field-theoretical expression for the low field magnetization, per site, $m(h)$ of the spin-1/2 Heisenberg chain [16]

$$m \rightarrow \frac{h}{2\pi v}[1 + \frac{1}{2\ln(J/h)}], \tag{11}$$

then we can estimate the critical magnetization $m_c \simeq 7.29 \times 10^{-4}$ at $h_c$. Here, spin wave $v = \pi J/2$, and the predominant exchange $J = 20.5$ K. Scaling dimensions of operators $N^+$ and $N^z$ are $\Delta_1$ and $\Delta_2$, respectively, with $\Delta_1 = \pi R^2$, $\Delta_2 = \pi/\beta^2$. In the limit of small magnetization, the parameter $2\pi R^2$ is well fitted [10] by

$$2\pi R^2 = 1 - \frac{1}{2\ln(M_0/m)}, \tag{12}$$

where $M_0 = \sqrt{8/(\pi e)}$. Therefore, the two scaling dimensions are modified by the magnetization in opposite ways,

$$\Delta_1 = \frac{1}{2}[1 - \frac{1}{2\ln(M_0/m)}], \quad \Delta_2 = \frac{1}{2[1-\frac{1}{2\ln(M_0/m)}]}. \tag{13}$$

These relations are illustrated in Fig. 6.

We now use ordering temperatures of the two competing orders, $T_{\rm sdw}(\Delta_2, J'/J)$ in Eq. (9) and $T_{\rm coneNN}(\Delta_1, J'/J)$ in Eq. (10), together with Eq. (13) and experimental value of $m_c$ in order to find the ratio $J_p = J'/J$. The critical magnetization $m_c$ satisfies the condition,

$$T_{\rm sdw}(m_c, J_p) = T_{\rm coneNN}(m_c, J_p). \tag{14}$$

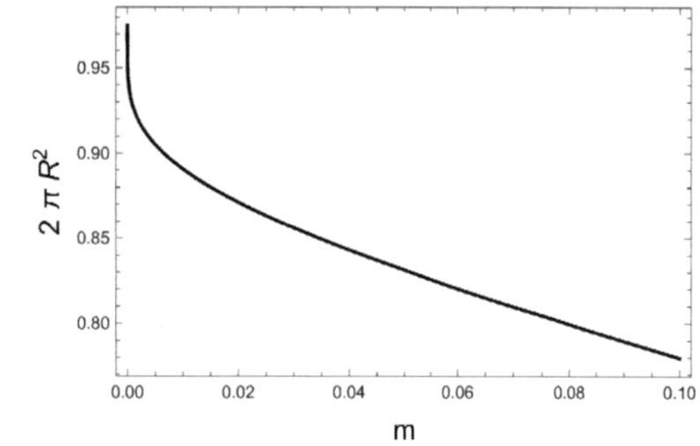


**Figure 5.** Parameter $2\pi R^2$ as a function of magnetization $m$, in the limit of small $m$, illustrating Eq. (12).

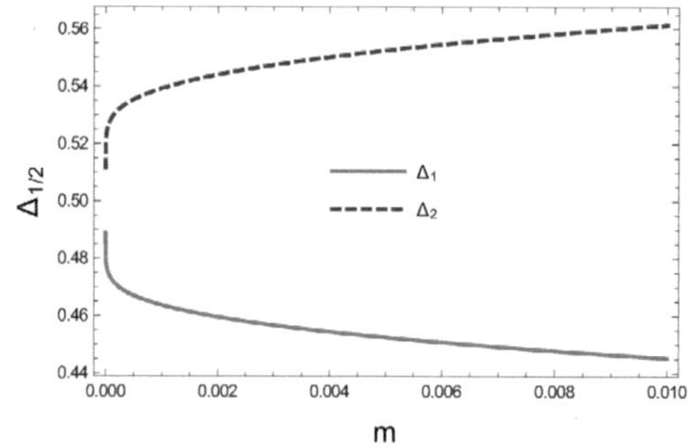


**Figure 6.** Two scaling dimensions $\Delta_{1/2}$ as a function of magnetization $m$, in the limit of small $m$, illustrating Eq. (13).

Fig. 7 shows that in the vicinity of the $m_c$, which is denoted by the red dot in the plot, $J_p$ varies linearly with magnetization $m$. We find $J_p = 0.01024$, which results in the inter-chain exchange $\mathcal{J}'_{\rm Br} \approx 0.2$ K. The curves for $T_{\rm sdw}$ and $T_{\rm coneNN}$ intersect at $T_0 \equiv T_{\rm sdw}(m_c, \mathcal{J}'_{\rm Br}) = T_{\rm coneNN}(m_c, \mathcal{J}'_{\rm Br}) \simeq 77$ mK.

With $\mathcal{J}'_{\rm Br} = 0.2$ K, we compute the zero field ordering temperature of SDW order, $T_{\rm sdw}|_{h=0} \approx 112.12$ mK, which agrees with the zero field specific heat anomaly at 100 mK [1].

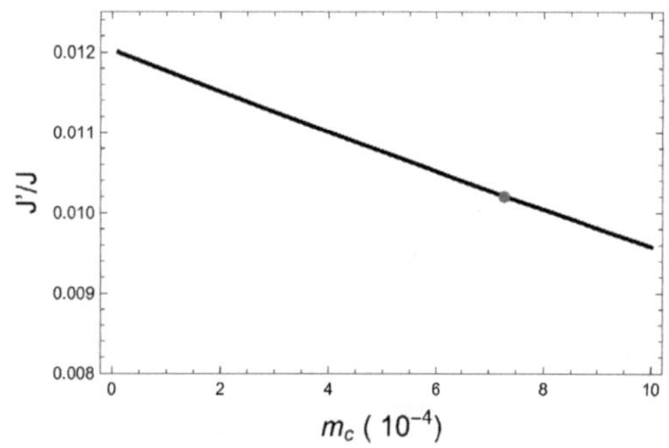


**Figure 7.** Relation between ratio $J_p = J'/J$ and magnetization $m$ near $m_c$. The coordinate of the red dot is $(m_c, J_p) = (7.29, 0.01024)$, which gives the inter-chain exchange for $K_2CuSO_4Br_2$ as $\mathcal{J}'_{\rm Br} \approx 0.2$ K.

## 5. Order parameters at $T = 0$

Here we propose to study the magnetic orders in more details by calculating the associate order parameters, even though experimental attempts to measure them, via neutron scattering and muon-spin spectroscopy, remain inconclusive for now [3]. The spin configuration is determined by the relative ordering of $\theta$ (for cone and coneNN states) and $\phi$ (for SDW) phases on neighboring chains, as well as by the magnitude of the local staggered magnetization, $\Psi$. The former is described in Appendix B, while the latter is calculated in Appendix D. For example, commensurate cone state is characterized by

$$\mathbf{S_{x,y}} = (0, 0, M) + (-1)^{x+y}|\Psi_{\rm cone}|(-\sin[\theta_0],\ \cos[\theta_0],\ 0). \tag{15}$$

The results of calculation in Appendix D are presented in Fig. 8 and 9. Comparing the two figures, we notice the order parameter has smaller magnitude in Br-compound, due to the stronger DM interaction which frustrates the system more. Also, cone-like orders are enhanced by magnetic field, while SDW order is monotonically suppressed by it.

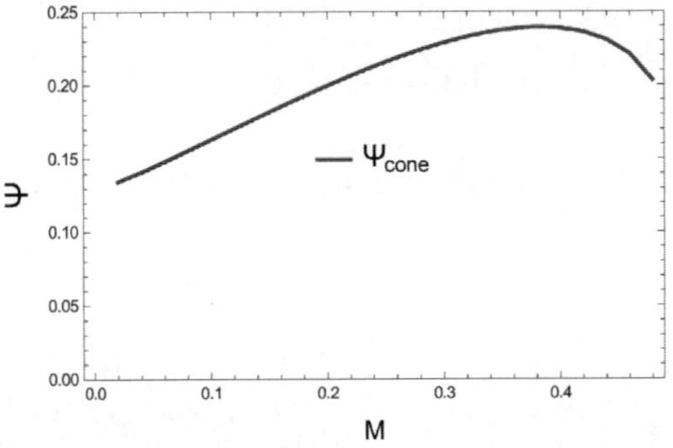


**Figure 8.** Order parameter of cone ($\Psi_{\text{cone}}$, blue) in $K_2CuSO_4Cl_2$, where $J'/J = 0.024$ and $D/J' = 0.55$, and $\Psi_{\text{cone}}$ is enhanced by field.

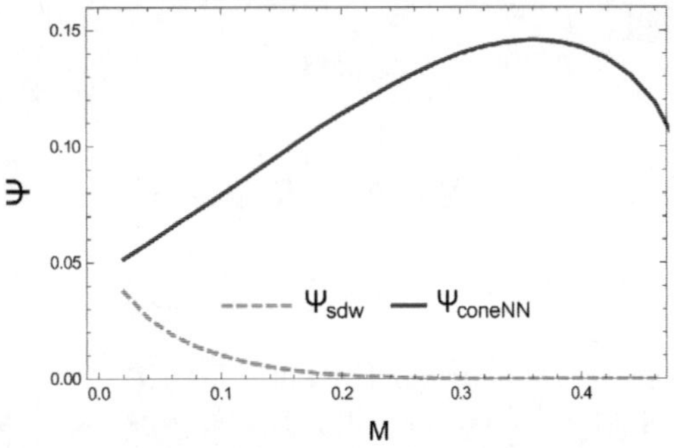


**Figure 9.** Order parameters of SDW ($\Psi_{\text{sdw}}$, orange dashed line) and coneNN ($\Psi_{\text{coneNN}}$, blue) in Br-compound, where $J'/J = 0.01$ and $D/J' = 1.4$.

## 6. Conclusion

We have studied effect of weak exchange $J'$ between antiferromagnetic Heisenberg spin-1/2 chains on the phase diagrams of quasi-one-dimensional materials $K_2CuSO_4Cl_2$ and $K_2CuSO_4Br_2$ [1, 2], subject to the uniform but *staggered between chains* Dzyaloshinskii-Moriya (DM) interaction and external magnetic field. With the help of chain mean field calculation, $J'$ can be determined quantitatively, and our results, denoted by $\mathcal{J}'$, are summarized in Table 1. For spin chains with strong DM interaction ($K_2CuSO_4Br_2$), one can extract value of $J'$ directly from the critical field corresponding to the phase transition between the spin density wave to the coneNN states. The order parameter at zero temperature for each state is calculated as well.

### Acknowledgments

We would like to thank A. Zheludev, M. Hälg and K. Povarov for detailed discussions of the experiments. This work was supported by the award from the National Science Foundation NSF DMR-1507054.

## Appendix A. Bosonization

In the low-energy continuum limit the spin operator is represented by [11],

$$\mathbf{S}(x) \to \mathbf{J}_L(x) + \mathbf{J}_R(x) + (-1)^{x/a}\mathbf{N}(x), \tag{A.1}$$

with $a$ is the lattice spacing, and continuous space coordinate is introduced via $x = na$, with $n$ an integer. $\mathbf{J}_L(x)$ and $\mathbf{J}_R(x)$, are the uniform left and right spin currents, and $\mathbf{N}(x)$ is the staggered magnetization. These fields can be conveniently expressed in terms of abelian bosonic fields $(\phi, \theta)$,

$$J_R^+ = \frac{1}{2\pi a} e^{-i\sqrt{2\pi}(\phi-\theta)},\ J_L^+ = \frac{1}{2\pi a} e^{i\sqrt{2\pi}(\phi+\theta)},\ J_R^z = \frac{\partial_x\phi - \partial_x\theta}{2\sqrt{2\pi}},\ J_L^z = \frac{\partial_x\phi + \partial_x\theta}{2\sqrt{2\pi}}. \tag{A.2}$$

$$\mathbf{N}(x) = \Big( - A_3 \sin[\beta\theta],\ A_3 \cos[\beta\theta],\ -A_1 \sin[\frac{2\pi}{\beta}\phi]\Big). \tag{A.3}$$

Then Hamiltonian in Eq. (A.4) can be expressed,

$$\mathcal{H} = \sum_y [\mathcal{H}_0 + \mathcal{V} + \mathcal{H}_{\text{bs}} + \mathcal{H}_{\text{inter}}], \tag{A.4}$$

where,

$$\mathcal{H}_0 = \frac{2\pi v}{3} \int dx(\mathbf{J_R} \cdot \mathbf{J_R} + \mathbf{J_L} \cdot \mathbf{J_L}) = \frac{v}{2} \int dx[(\partial_x \phi)^2 + (\partial_x \theta)^2], \tag{A.5}$$

$$\mathcal{V} = \mathcal{H}_Z + \mathcal{H}_{DM} = -h_z \int dx(J_R^z + J_L^z) + (-1)^y D \int dx(J_R^z - J_L^z), \tag{A.6}$$

$$\mathcal{H}_{bs} = -g_{bs} \int dx[J_R^x J_L^x + J_R^y J_L^y + (1+\lambda) J_R^z J_L^z], \quad \mathcal{H}_{inter} = J' \int dx \mathbf{N}_y \cdot \mathbf{N}_{y+1}, \tag{A.7}$$

where $v \simeq J\pi a/2$ is the spin velocity, $\lambda = c' D^2/J^2$ describes an Ising-like anisotropy induced by DM interaction [13]. $\mathcal{V}$ contains the last two terms of Eq. (1), it collects all vector-like perturbations of the bare chain Hamiltonian $\mathcal{H}_0$. $\mathcal{H}_Z$ and $\mathcal{H}_{DM}$ are the Zeeman and DM interactions, respectively. $\mathcal{H}_{bs}$ describes residual backscattering interaction between right- and left-moving spin modes of the chain, its coupling is estimated as $g_{bs} \approx 0.23 \times (2\pi v)$, see Ref. [13] for details. The constant $c' = (2\sqrt{2}v/g_{bs})^2$ is about 3.83 from the Bethe ansatz solution, see (B2) in Ref. [13]. The inter-chain interaction is described by $\mathcal{H}_{inter}$, in which we kept the most relevant, in renormalization group sense, contribution, $\mathbf{S}_{x,y} \cdot \mathbf{S}_{x,y+1} \to \mathbf{N}_{x,y} \cdot \mathbf{N}_{x,y+1}$. Evidently, the linear terms in $\mathcal{V}$ can be *absorbed* into $\mathcal{H}_0$ by shifting fields $\phi$ and $\theta$ appropriately,

$$\phi \to \phi + \frac{t_\phi}{\sqrt{2\pi}} x, \qquad t_\phi \equiv \frac{h}{v}, \tag{A.8}$$

$$\theta \to \theta + (-1)^y \frac{t_\theta}{\sqrt{2\pi}} x, \qquad t_\theta \equiv \frac{D}{v}. \tag{A.9}$$

As a result of the shifts, the spin currents and the staggered magnetization are modified as

$$J_R^+ \to J_R^+ e^{-i(t_\phi - t_\theta^y)x}, \; J_L^+ \to J_L^+ e^{i(t_\phi + t_\theta^y)x}, \; J_R^z \to J_R^z + \frac{(t_\phi - t_\theta^y)}{4\pi}, \; J_L^z \to J_L^z + \frac{(t_\phi + t_\theta^y)}{4\pi}, \tag{A.10}$$

$$N^+ \to N^+ e^{it_\theta^y x}, \quad N^z \to -A_1 \sin(\frac{2\pi}{\beta}\phi + t_\phi x). \tag{A.11}$$

Here $t_\theta^y \equiv (-1)^y t_\theta$ depends on the parity of the chain index $y$. Notice that the shift introduces oscillating position-dependent factors to transverse components of $\mathbf{J}$ and $\mathbf{N}$. The Hamiltonian now reads as Eq. (4). The complete form of $\mathcal{H}_{sdw}$ is,

$$\begin{aligned} \mathcal{H}_{sdw} &= J' A_1^2 \int dx \sin[\frac{2\pi}{\beta}\phi_y + t_\phi x] \sin[\frac{2\pi}{\beta}\phi_{y+1} + t_\phi x] \\ &= \frac{1}{2} J' A_1^2 \int dx \Big\{ \cos[\frac{2\pi}{\beta}(\phi_y - \phi_{y+1})] - \cos[\frac{2\pi}{\beta}(\phi_y + \phi_{y+1}) + 2t_\phi x] \Big\}. \end{aligned} \tag{A.12}$$

The effect of position-dependent phase $t_\phi x$ is to induce sign-changing oscillations on the spatial scale $\propto 1/t_\phi \sim J/h$. The corresponding RG scale is $\ell_\phi \sim \ln(J/h)$. This needs to be compared with the RG scale $\ell_{inter} \sim \ln(J/J')$ on which $\mathcal{H}_{sdw}$ reaches strong coupling. For $\ell_{inter} \ll \ell_\phi$ ($h \ll J'$) the oscillations are not important and SDW order is commensurate. In the opposite limit $\ell_{inter} \gg \ell_\phi$ ($h \gg J'$) the oscillations are frequent and wash out the second term in (A.12). This corresponds to the development of incommensurate SDW. The commensurate-incommensurate change takes place at $h \sim J'$. Here we assume that magnetic field $h_c$, at which SDW-coneNN transition takes place (see Sec. 4), satisfies $h_c \geq J'$ condition and use the reduced form of $\mathcal{H}_{sdw}$, first term in (A.12) and Eq. (4). The obtained estimate for the interchain $J' \approx 0.2$ K is found to be consistent with the made assumption.

## Appendix B. Spin configurations

The spin operator has two parts,

$$\mathbf{S_{x,y}} = \mathbf{M} + (-1)^x \mathbf{N}_y(x), \tag{B.1}$$

with $\mathbf{M}$ is magnetization along the magnetic field, and $\mathbf{N_y(x)}$ is the staggered magnetization on site $(x, y)$. We have three competing interactions $\mathcal{H}_{\text{cone}}$, $\mathcal{H}_{\text{sdw}}$ and $\mathcal{H}_{\text{NN}}$ as in Eq. (4) and (6). When the dominant interaction is $\mathcal{H}_{\text{cone}}$, and the ground state is commensurate cone phase.

$$\mathbf{N}_y(x) = (-1)^y |\Psi_{\text{cone}}|(-\sin[\theta_0],\ \cos[\theta_0],\ 0). \tag{B.2}$$

This cone order is illustrated in Fig. B1. When the most relevant interaction is $\mathcal{H}_{\text{sdw}}$, the ground state is collinear SDW state. The system will be ordered in $\phi$ field, with $\phi_y = \phi_0 + \sqrt{\frac{\pi}{2}} y + \frac{hx}{\sqrt{2\pi}v}$, and staggered magnetization is

$$\mathbf{N}_y(x) = (-1)^y |\Psi_{\text{sdw}}|(0,\ 0,\ \sin[\sqrt{2\pi}\phi_0 + \frac{hx}{v}]). \tag{B.3}$$

If $\mathcal{H}_{\text{NN}}$ is the most relevant interchain interaction, we obtain cone-like order in the plane perpendicular to the external magnetic field, similar to $\mathcal{H}_{\text{cone}}$ case. The order corresponds to $\theta_y = \theta_0 + (-1)^y \frac{Dx}{\sqrt{2\pi}v}$. (In principle, the constant phase $\theta_0$ can be different for *even* and *odd* chains – it remains unclear how these two subsystems couple to each other in the regime of strong DM interaction.) As a result, the staggered magnetization reads

$$\mathbf{N}_y(x) = |\Psi_{\text{coneNN}}|(-\sin[\sqrt{2\pi}\theta_0 + (-1)^y Dx/v],\ \cos[\sqrt{2\pi}\theta_0 + (-1)^y Dx/v],\ 0). \tag{B.4}$$

This order is shown in Fig.2. Different from order in (B.2), coneNN is spiraling in the transverse $x - y$ plane.

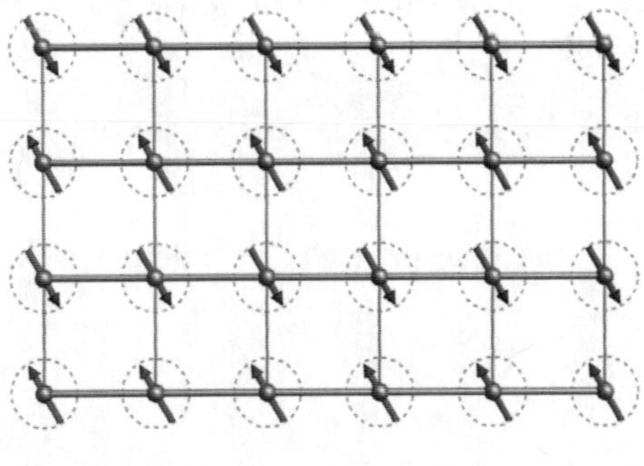

**Figure B1.** Staggered magnetization of the cone state, illustrated for a field $\mathbf{h} \parallel \mathbf{D}$, where all spins are ordered in the transverse plane. This order is commensurate. Spins pointing to opposite directions, reflecting the anti-ferromagnetic interaction between neighboring chains.

## Appendix C. Chain mean field approximation

For an interaction,

$$\mathcal{H}_y = \mathcal{H}_0 - c \int dx \cos[\beta\theta_y] \cos[\beta\theta_{y+1}], \tag{C.1}$$

with the approximation of chain mean field(CMF), the spin (staggered magnetization part) takes an average value

$$\tilde{\psi} = \langle \cos[\beta\theta_y] \rangle. \tag{C.2}$$

Then the Hamiltonian (C.1) reduces to,

$$\mathcal{H}_y = \mathcal{H}_0 - 2c\tilde{\psi} \int dx \cos[\beta\theta_y], \tag{C.3}$$

where the factor of 2 arises from two neighboring chains. Eq. (C.3) is just the sine-Gordon Hamiltonian, to which three interchain interactions in Eq. (4) and (6) along with $\mathcal{H}_0$ in Eq. (A.5) can reduce. We can rearrange $\mathcal{H}_{\text{cone}}$ in Eq. (4), by shifting the fields $\theta$,

$$\theta_y \to \tilde{\theta}_y + (-1)^y \pi/(2\beta) - (-1)^y t_\theta x, \tag{C.4}$$

then $\mathcal{H}_{\text{cone}}$ changes sign and transforms to,

$$\mathcal{H}_{\text{cone}} = -2c_1\tilde{\psi}_1 \int dx \cos[\beta\tilde{\theta}_y], \tag{C.5}$$

with $\tilde{\psi}_1 = \langle\cos[\beta\tilde{\theta}]\rangle$ and $\cos\beta\tilde{\theta} \sim \cos[\beta\theta + (-1)^y t_\theta x]$ is the $x$-component of staggered magnetization in Eq. (A.11). Similar procedure (but shift is independent of position) can be applied to $\mathcal{H}_{\text{sdw}}$ in Eq. (4) and $\mathcal{H}_{\text{NN}}$ in Eq. (6),

$$\mathcal{H}_{\text{sdw}} = -2c_2\tilde{\psi}_2 \int dx[\cos\frac{2\pi}{\beta}\phi_y], \quad \mathcal{H}_{\text{NN}} = -2c_3\tilde{\psi}_3 \int dx \cos[\beta\theta_y], \tag{C.6}$$

where $\tilde{\psi}_2 = \langle\cos[2\pi\phi/\beta]\rangle$, $\tilde{\psi}_3 = \langle\cos[\beta\theta]\rangle$.

To determine the critical temperature, we can expand Eq. (C.3), with self-consistent condition,

$$\frac{1}{2c} = \chi(q=0, \omega_n = 0; T_c) \quad = \int dx \int_0^{1/T_c} d\tau e^{iqx+\omega_n\tau}\langle O(x,\tau)O(0,0)\rangle_0 \tag{C.7}$$

$\chi(q, \omega_n; T)$ is momentum and frequency dependent susceptibility at finite temperature $T$. Here the operators in consideration are,

$$O = \cos(\sqrt{4\pi\Delta_1}\theta) \quad \text{or} \quad O = \cos(\sqrt{4\pi\Delta_2}\phi). \tag{C.8}$$

With two scaling dimensions in Table. 2, $\Delta_1 = \pi R^2$ and $\Delta_2 = \pi/\beta^2$. Here we follow the standard calculation in Ref. [10], where Eq. (D.55) gives the expression of static susceptibility,

$$\chi(q=0, \omega_n=0; T_c) = \frac{\pi u\alpha^2}{2}\Big[(2\pi T\alpha)^{2\Delta-2}\frac{\Gamma(1-\Delta)\Gamma(\Delta/2)^2}{\Gamma(\Delta)\Gamma(1-\Delta/2)^2} - \frac{\Gamma(\Delta-1/2)}{\sqrt{\pi}(1-\Delta)\Gamma(\Delta)}\Big]. \tag{C.9}$$

Here, $\Delta$ is either $\Delta_1$ or $\Delta_2$, and second term removes the non-physical divergence as $\Delta$ approaching to 1.

## Appendix D. Order parameter at $T = 0$

For a sine-Gordon model in Eq. (C.3), its action

$$S_{\text{sG}} = \int dx dy\Big(\frac{1}{2}(\partial_x\theta)^2 + \frac{1}{2}(\partial_y\theta)^2 - 2\mu\cos[\beta\theta]\Big). \tag{D.1}$$

here, $\mu = c\langle\cos[\beta\theta]\rangle/v$, and $\tau = y/v$. According to Ref. [10, 17], we have the expression for $\tilde{\psi} \equiv \langle\cos\beta\theta\rangle$ as a function of magnetization $M$,

$$\tilde{\psi}(M) = \Big[(\frac{c}{v})^{\beta'^2}\sigma'(M)^{1-\beta'^2}\Big]^{1/(1-2\beta'^2)}, \tag{D.2}$$

where $\beta' = \beta/\sqrt{8\pi}$, and

$$\sigma'(M) = \frac{\tan[\pi\xi/2]}{2\pi(1-\beta'^2)}\left[\frac{\Gamma(\frac{\xi}{2})}{\Gamma(\frac{1+\xi}{2})}\right]^2\Big[\frac{\pi\Gamma(1-\beta'^2)}{\Gamma(\beta'^2)}\Big]^{1/(1-\beta'^2)}, \quad \xi = \frac{\beta'^2}{1-\beta'^2} = \frac{\beta^2}{8\pi-\beta^2}, \tag{D.3}$$

Eq. (D.2) is a general form of order parameter for sine-Gordon model. The three interactions in consideration are Eq. (C.4) and (C.6), with $\beta = 2\pi R$, and their corresponding parameter $\beta'$ are,

$$\beta'_{1/3} = \Delta_1/2, \quad \beta'_2 = \Delta_2/2, \tag{D.4}$$

where $\beta'_{1,2,3}$ are associated with $\tilde{\psi}_{1,2,3}$. Now we can compute the order parameters for two materials $K_2CuSO_4Cl_2$ and $K_2CuSO_4Br_2$ with their exchange constants in Table 1. For $K_2CuSO_4Cl_2$, there is single cone with order parameter,

$$\Psi_{\text{cone}} = A_3\Big[(\frac{c_1}{v})^{\Delta_1}\sigma'(M)^{2-\Delta_1}\Big]^{1/(2-2\Delta_1)}. \tag{D.5}$$

$\Psi_{\text{cone}}$ is shown in Fig. 8. For $K_2CuSO_4Br_2$, two order parameters need to be considered,

$$\Psi_{\text{sdw}} = A_1\Big[(\frac{c_2}{v})^{\Delta_2}\sigma'(M)^{2-\Delta_2}\Big]^{1/(2-2\Delta_2)}, \quad \Psi_{\text{coneNN}} = A_3\Big[(\frac{c_3}{v})^{\Delta_1}\sigma'(M)^{2-\Delta_1}\Big]^{1/(2-2\Delta_1)} \tag{D.6}$$

and they are shown in Fig. 9.

## References

[1] Hälg M, Lorenz W E A, Povarov K Y, Månsson M, Skourski Y and Zheludev A 2014 *Phys. Rev. B* **90**(17) 174413
[2] Smirnov A I, Soldatov T A, Povarov K Y, Hälg M, Lorenz W E A and Zheludev A 2015 *Phys. Rev. B* **92**(13) 134417
[3] Hälg M 2015 *Quantum criticality, universality and scaling in organometallic spin-chain compounds* Ph.D. thesis ETH-ZÃijrich
[4] Jin W and Starykh O A 2016 Phase diagram of weakly coupled heisenberg chains subject to a uniform dzyaloshinskii-moriya interaction *In preparation*
[5] Kimchi I and Coldea R 2016 *ArXiv e-prints* (*Preprint* 1606.05356)
[6] Johnston D C, Kremer R K, Troyer M, Wang X, Klümper A, Bud'ko S L, Panchula A F and Canfield P C 2000 *Phys. Rev. B* **61**(14) 9558–9606
[7] Müller G, Thomas H, Beck H and Bonner J C 1981 *Phys. Rev. B* **24**(3) 1429–1467
[8] Povarov K Y, Smirnov A I, Starykh O A, Petrov S V and Shapiro A Y 2011 *Phys. Rev. Lett.* **107**(3) 037204
[9] Yasuda C, Todo S, Hukushima K, Alet F, Keller M, Troyer M and Takayama H 2005 *Phys. Rev. Lett.* **94**(21) 217201
[10] Starykh O A, Katsura H and Balents L 2010 *Phys. Rev. B* **82**(1) 014421
[11] Gangadharaiah S, Sun J and Starykh O A 2008 *Phys. Rev. B* **78**(5) 054436
[12] Schnyder A P, Starykh O A and Balents L 2008 *Phys. Rev. B* **78**(17) 174420
[13] Garate I and Affleck I 2010 *Phys. Rev. B* **81**(14) 144419
[14] Essler F H L, Furusaki A and Hikihara T 2003 *Phys. Rev. B* **68**(6) 064410
[15] Starykh O A and Balents L 2007 *Phys. Rev. Lett.* **98**(7) 077205
[16] Affleck I and Oshikawa M 1999 *Phys. Rev. B* **60**(2) 1038–1056
[17] Lukyanov S and Zamolodchikov A 1997 *Nuclear Physics B* **493**(3) 571 – 587